GEOGRAPHY OF UNDERDEVELOPMENT

To
My beloved wife Girija

About the Author

K. V. SUNDARAM, M.A., D. Litt., is a well known Geographer-Regional Planner, now with the Indian Planning Commission as Deputy Adviser in charge of multi-level planning. He has been variously Consultant to the United Nations Organisations, such as the United Nations, Centre for Regional Development, Nagoya (Japan) and ESCAP, Bangkok. He was assisted by a Nuffield grant during 1978-79 and spent two terms of residence in the Centre of South Asian Studies, University of Cambridge (UK), when this work was undertaken. His many publications include : *Regional Development Planning in India: A New Strategy; Urban and Regional Planning in India; Multi-level Planning and Integrated Rural Development in India;* and *Rural Area Development: Perspectives and Approaches.*

GEOGRAPHY OF UNDERDEVELOPMENT

The Spatial Dynamics of Underdevelopment

K.V. SUNDARAM

CONCEPT PUBLISHING COMPANY PVT. LTD.
NEW DELHI-110059

ISBN: 978-93-5594-878-6

Edition 2024

Published and Printed by

Since 1974

Concept Publishing Company Pvt. Ltd.
A/15-16, Commercial Block, Mohan Garden,
New Delhi-110059 (India)
T: 91-11-41101460
E: publishing@conceptpub.com, *W*: www.conceptpub.com

FOREWORD

It is a great privilege, and gives me much pleasure, to be asked to write a Foreword for Dr. K.V. Sundaram's *magnum opus*, his *Geography of Underdevelopment : The Spatial Dynamics of Underdevelopment.*

I first met Dr. Sundaram at the First Indo-British Geographical Seminar held in Delhi in 1972 (having been delayed for a year by the Bangladesh War of 1971). He impressed me, as he impressed other British participants previously unaware of him and his work, as a man of great ability destined to make a great contribution to geography, to the study of development, and to his country. His subsequent career, particularly since he joined the Planning Commission with special responsibility for multi-level planning, has not disappointed us.

Dr. Sundaram visited the Centre of South Asian Studies in the University of Cambridge in March, 1974. With funds generously provided by the Nuffield Foundation we were able to arrange for him to pay two longer visits, each of about three months' duration, from March to May, 1978 and from April to June, 1979. My colleagues and I were able to get to know him more thoroughly, and our appreciation of him deepened. As he has been so kind as to make clear in his Introduction, a substantial part of the work for this impressive and well-researched book was done during these visits; and we were glad to provide him with some peace and quiet away from the hurly-burly of official life. He in his turn made a notable contribution to the life and work of the Centre.

Dr. Sundaram's book is to be warmly commended to geographers of development and of South and Southeast Asia (from which he draws his case studies); to students of development in other disciplines (some of whom are still insufficiently aware of geographical and spatial dimensions in their work); and to planners, politicians and administrators of development. I hope that it will have the very considerable intellectual and practical impact that it deserves.

Centre of South Asian Studies
University of Cambridge
ENGLAND

B. H. FARMER

27 January 1982

PREFACE

THE BEGINNING of this work may be traced to 1973-74, when, as a Consultant to the United Nations, I happened to visit a number of South and South-East Asian countries. These visits enabled me to gain some visual perception of the nature of developments going on in these countries. I was particularly struck by the serious imbalances of various kinds persisting in the sub-national areas of these countries and the struggle of the governments to cope with these problems. Since then, the idea of studying the developmental problems and policies in these countries from the point of view of sub-national integration grew in me. During my several visits abroad, a lot of data and literature were collected. I had also opportunities to discuss these problems in depth with several scholars from these countries. But in the daily tasks of governmental work, I could hardly find the time for reading, reflection and writing. This stalemate was, however, broken in 1978 when the Nuffield Foundation in England conferred an award on me with generous funds to support my stay in Cambridge. The Centre of South Asian Studies, University of Cambridge (England) came forward to provide me the physical facilities necessary to carry out the work. I spent two terms of residence in Cambridge during the summers of 1978 and 1979, when the bulk of the writing of this work was done.

My fundamental acknowledgements go to the Nuffield Foundation in England and the Centre for South Asian Studies, Cambridge who made this work possible.

I will be failing in my duty if I do not record my deep sense of gratitude to Dr. B.H. Farmer, Director of the Centre of South Asian Studies, Cambridge for all the guidance, support and assistance rendered to me during the preparation of this work. Dr. Farmer also read through my manuscript in part and offered very helpful suggestions. The St. John's College,

Cambridge gave me the privilege of dining at the high table. The staff of the Centre of South Asian Studies, Mr. A.H.N. Richards, Mrs. Pauline Doncaster and Mrs. C. Brown showed me exceptional kindness and facilitated my task in various ways. I am grateful to all of them for their cooperation and assistance. My stay in Cambridge also enabled me to interact with a number of scholars in England and discuss my work. Among them, special mention may be made of Prof. Michael Chisholm and Dr. Graham Chapman of the Department of Geography, University of Cambridge, Dr. Sudhir Wanmali of the Centre of South Asian Studies, Prof. Charles Elliott, Dr. Ray Bromley and Dr. H.R.J. Davies of the University College of Swansea,, Prof. David Thomas, Professor of Geography, University of Birmingham and Dr. J.R. Watsom and Dr. Donald Curtis of the Institute of Local Government Studies, University of Birmingham. In May 1979, I was invited by a number of scholars in Europe and U.S.A. working on this and allied themes to give seminars and lectures. I visited France, the Netherlands, West Germany, Denmark and U.S.A. and exchanged views with a number of scholars. Particular mention may be made of my discussions with Prof. J. Hinderink and Dr. George Tempelman of Geografisch Institut, University of Utrecht (Netherlands), Prof. Guy Lasserre of the Centre for Tropical Geography, University of Bordeaux (France), Dr. H.S. Muller of the School of Economics and Business Administration, University of Copenhagen (Denmark), Prof. Dr. Martin Pfaff of the Department of Macro-Economics, University of Augsburg (West Germany), Prof. Dr. F. Stang of Technical University, Aachen (West Germany), Dr. H.C. Reiger and Prof. Dr. Bruno Knall of the South Asia Institute, University of Heidelberg (West Germany). In U.S.A., I had discussions with Prof. T.R. Lakshmanan, and Dr. Lata Chatterjee of the Department of Geography, Boston University, Prof. Rhoads Murphey, Department of Geography in the University of Michigan, Ann Arbor and Prof. Gustav Papanek, Chairman, Department of Economics, Boston University. All these discussions and seminars helped me a great deal in understanding the various view-points and to firm up my own conclusions on the controversial theme discussed in the present work. I benefitted greatly from the comments of these scholars. The many imperfections in this work are my sole responsibility.

In India, the Planning Commission (Government of India) was kind enough to grant me study leave to enable me to avail of the Nuffield Fellowship. My profound thanks are due to Dr. Ajit Mazoomdar and Mr. P.H. Vaishnav, Secretary and Joint Secretary respectively in the Planning Commission, who gave me all the encouragement and released me for the period of the fellowship.

Dr. R.P. Misra, Deputy Director, United Nations Centre for Regional Development, Nagoya (Japan) read through the final manuscript. I am grateful to him for spending many of his valuable hours and offering some very useful suggestions. I am also grateful to my father-in-law Mr. S.V. Subramanian who read through the manuscript and suggested some changes for improving the literary presentation of this work. The views expressed in this book are, however, mine and may not be attributed to the organization to which I belong.

Lastly, I would like to add that this work could not have been completed but for the understanding and cooperation shown by my wife, Girija to whom I affectionately dedicate this book.

NEW DELHI K.V. SUNDARAM

9 February 1982

CONTENTS

PART ONE

THE CONCEPTUAL FRAMEWORK

CHAPTER 1

INTRODUCTION

And thou wert still a hope, a love,
Still longed for, never seen

—*William Wordsworth*

IN RECENT years, planning for the developing countries has shifted its emphasis from a growth-oriented paradigm into one in which the concern for a significant narrowing of disparities in incomes or levels of living (that is distributional equity) has become increasingly important. This concern for distributional equity has many dimensions—sectoral, spatial, social and political. It is therefore increasingly being realized now that the policies for developing countries should be informed by a multi-dimensional perspective. Case studies and other historical and contemporary evidence about developing countries have also demonstrated the need for a development policy in developing countries to directly focus on the poor and marginal groups in society and recognise the worth of many traditional values, structures and processes in relation to the development process. Otherwise the development process cannot achieve its ultimate goal of human welfare.[1]

This shift in development emphasis is very well reflected in the UN Resolution for the "Second Development Decade" adopted in 1970, which emphasises the need to reduce "regional, sectoral and social disparities", and in a series of subsequent resolutions urging member governments to utilise a "unified approach" in the preparation of their development plans.[2]

The problems of social and economic inequality have constituted a subject of discussion for quite some time. During the second half of the 19th century, this issue of the unequal distribution of income—considered synonymous with inequality in general—received the attention of economists and sociologists, such as John Stuart Mill, Karl Marx and De Tocqueville. In the 20th century, and particularly during the period preceding World War II, it was Keynes who exercised considerable influence on the ideas about income inequality and economic growth.[3] These writers, however, confined their attention to the inequality problems of the developed industrialised countries. The problems of the developing countries did not receive much attention during this period.

Since the middle of this century, increasing attention has been bestowed on the developing countries focussing on socio-economic inequality in general and on income inequality in particular. Studies have been undertaken focussing on the quantitative aspects of income inequality, on changes in income inequality over time and on income distribution at various levels of economic development.

The cross-country studies undertaken by Kuznets, Williamson, Adelman and Morris, Chenery and Ahluwalia[4] may be particularly mentioned in this context. Such cross-country analyses have provided useful summary pictures of the overall degree of inequality among nations and helped in the classification of countries into certain types. But they do not tell us much about the underlying socio-economic structures causing inequality. These structures may vary widely between countries and yet produce the same overall degree of inequality. While such studies have undoubtedly promoted the understanding of development processes in general and also provided some policy directions to national governments and world bodies concerned with development issues, they have not provided any *firm* basis for policy formulation towards inter-regional reduction in disparities. This is because such studies have, by and large, depended on a highly aggregative analysis. Comparatively few studies have been made highlighting the intra-country differences in income and other socio-economic characteristics of development in the developing countries. The reason is partly data limitations. The situation however is rapidly changing. A large number of surveys have been carried out in several underdeveloped countries in recent years and results from these surveys are increasingly being used in the analyses of development problems.[5]

The emphasis of the present study will be on the various types of inequalities that occur *within* a country and a further orientation will be on underdevelopment viewed in its spatial dimension.

Underdevelopment is a relative concept. Basically, it stems from inequalities of various kinds[6]—geographical, developmental and social—occurring in their various combinations and groupings, rendering the spatial structure of the developing countries a complicated tapestry. To this syndrome of inequalities, I have applied the term spatial inequality. Obviously, the concept of spatial inequality, as understood here, needs some elaboration.

It is common knowledge that in every country, factor and resource endowments are unequally distributed in space. This is Nature's dispensation. We have to simply take it as a datum. We can do precious little about it. Even then, through our efforts that is through the application of science and technology, we can considerably enhance their marginal utility through efficient patterns of production, management and use. Thus the Nature imposed inequality is one type of inequality which gives rise to what we call a relatively static and fundamental inequality in space. We may call this by the term geographical inequalities. Such inequalities give rise to certain fundamental differences in economic opportunities among the different regions.

Upon this fundamental pattern of geographical inequalities, man's development efforts inscribe other patterns of inequality, caused by differential interventions in space. Such types of inequalities are caused by differences in growth rates of the indigenous supplies of factors of production, namely, land, labour and capital. These may be called developmental inequalities. This type of inequality is not a static one. Here we are dealing with a situation in which everything is dynamic and the growth rates are discordant.

Apart from the geographical and development inequalities, there are also the so-called social inequalities or better called socio-economic inequalities,[7] characterising the human resource endowment. This is particularly sharp in the developing countries. These are caused by class, group, ethnic, cultural and income differences among human beings, differences in opportunity structures and differences in access to development benefits that arise because of the exploitation of man over his brother man. Whereas geographical inequalities are present in a horizontal direction, developmental as well as socio-economic inequalities are present both in horizontal as well as vertical directions. It must be noted that geographical inequalities compounded by developmental inequalities and socio-economic inequalities impinge on space and manifest themselves in varying and complex patterns of concentration and dispersal of inequalities. This complex spatial configuration of inequality mix of both static and dynamic elements is what I have called by the term spatial inequality.

Thus within the rubric of spatial inequality, all the three kinds of inequalities are contained. It is therefore a comprehensive term. Sometimes, spatial inequality is narrowly interpreted to mean only those inequalities, wherein space or the attributes of space (for example, location, space relations and accessibility) enter as distinct variables contributing to differentiation or inequality.[8] Such a view tends to ignore the other dimensions namely, the developmental and socio-economic inequalities. In real life situations, it is difficult to isolate them because they are closely intertwined, complex and operate together as mutually reinforcing variables. We find that the elements of the one are usually reflected or woven into the other. It must also be noted that "many of the causes, consequences and manifestations of inequality are dominantly structural rather than spatial in nature".[9] Thus, by spatial inequality is meant the study of uneven development (differentiation of development) in space, which is characterised by a certain pattern of association of different variables (including space, environmental, developmental and socio-economic variables) in an area contributing to a syndrome of inequality. Such association may be causal or incidental or both. It is this broader definition of spatial inequality that has been adopted in this study.

Analysis of spatial inequality implies separating the various strands contributing to inequalities (both in their vertical and horizontal dimensions) and examining the role of each in relation to the other, and also in relation to economic development. Areal differences and their relationships will also constitute an important aspect of such analysis. In this manner, it will be possible to get a total understanding of the different patterns of underdevelopment, the complex reasons that have led to these patterns and the possible directions in which solutions could be devised towards bringing about a more even pattern of development.

The problem of spatial inequality is both a *universal* as well as a *unique* problem. It is universal in the sense that it is met with in both the developed as well as the developing countries. It is unique in the sense that each type of underdevelopment must be dealt with in its specific context for planning and development purposes. Even within a single country, the patterns of underdevelopmnet may vary. Thus there may be different types of backwardness calling for unique and specific treatment.* The macro-growth policies usually bypass their uniqueness or specificity in particular areas and have failed to capture all their problems. The recognition of the failure of such policies has prompted many developing countries to think of devising more appropriate approaches and policies for the development of their specific areas rooted in the empirical realities of the situation. Such area specific or regional policies, range from isolated and *ad hoc* approaches to more comprehensive measures, including area or regional planning. The effects of such area or regional policies in developing countries are neither fully known nor closely evaluated.

While some of these policies might have led to increased production and incomes in the backward areas in an aggregate sense, there is also evidence to suggest that they have not benefited the less mobile and less affluent population strata in these countries, as the relatively more advantaged groups, who have better access, always manage to corner the development benefits. Thus these policies have led to a variety of leakages of the positive effects of economic development. Further, some of these policies have always viewed backward area development as an extension of development from the core (that is,

*The concept of areal differentiation implies the recognition of such types in space.

from the advanced) to the peripheral areas; naturally, the transference of such development has followed the standards and value systems of the core or advanced areas. This importation has led to conflicts with the indigenous standards and value systems in the peripheral areas. Such an exogenous factor in backward area development has also had other negative consequences, such as development of an attitude of excessive external dependence and the reduction in self-reliance of the peripheral areas. These are some of the problems created by the development policies themselves in the peripheral areas, which call for appropriate correctives, adaptations and reorientation. In many recent discussions, the need for developing a realistic regional strategy that will have a stronger empirical foundation and will be directed at the specific problems of the lagging regions in the developing countries and promote a more self-reliant development in these countries has been stressed.[10]

The present work proposes to go into these various aspects of underdevelopment in five selected countries—Malaysia, Philippines, Indonesia, Thailand and India.[11] The objectives of this study are :

(i) to present a clear picture of the spatial inequalities characterising the development scene in these countries;
(ii) to examine their patterns, typologies and origins;
(iii) to identify the major regional policy instruments and "measures of regional closure" adopted by these countries;
(iv) to analyse the experience of policies implemented in these countries; and
(v) to outline the elements of coherent policy framework for the development of backward areas in the developing countries.

Among the five countries selected for this study, the Indian case is dealt with in a more comprehensive manner than the other countries for the simple reason that the author has had greater access to literature and been engaged in various investigations relating to this problem for more than a decade. In the case of other countries, the author has relied on extensive secondary sources—studies, data and information available in a somewhat fragmented form. An attempt has been made to present each case in a consistent manner so as to conform to a general design of analysis in which the emphasis is on setting out the dimensions of the problem of uneven development and on highlighting the relative effectiveness of different policies and strategies pursued to tackle the problems. While doing so, historical, descriptive and statistical methods of analysis have been combined to provide comprehensive insights and an attempt has been made to highlight the differing institutional frameworks and cultural values within which the different economies have attempted to tackle the problem of development disparities. The lack of uniformity in the data has precluded an attempt at a highly uniform and systematic empirical investigation of the problem in the different countries. Inevitably, this has left some issues of the problem only dimly perceived at present. Any multi-national comparative investigation of inter-regional disparities has to face such problems, although it must be agreed that the data base in individual countries is improving considerably.

Notwithstanding these shortcomings, the study has helped to identify the numerous factors which underlie the patterns of uneven development in these countries and to understand the nature of policy interventions that have been attempted to tackle the problem of spatial inequalities. At a time when the regional gap within the developing countries seems to be widening, rather than narrowing,* the need for understanding the pathology of underdevelopment in country-specific situations by social scientists is perhaps greater than ever before. In this context, the present work is offered as a modest contribution, both at the diagnostic and policy action levels on the one hand, and at the conceptual level on the other.

*Some evidence in this regard is presented in Chapter 2 of this volume.

NOTES AND REFERENCES

1. In keeping with this shift in thinking, scholarly inquiry, as well as governmental policy determination and implementation in developing countries are becoming increasingly oriented towards questions concerning the economic and social development of the so-called underdeveloped areas.

The following statement in an ECLA document is significant:
"The problem before the developing countries today is not just a question of raising the growth rates, but of laying the basic institutional foundations of a structure that will increase the social impact to development by absorbing labour in productive activities, ensuring a more equitable distribution of poverty and income and resolve the problems of sectoral imbalances and backward regions within countries." Economic Commission for Latin America (ECLA): Income Distribution in Latin America, New York, 1971.

2. See Resolution 1494 (XLVIII) of the Economic and Social Council, May 26, 1970, and General Assembly Resolution 2681 (XXV), December 11, 1970. Among the more recent is General Assembly Resolution 3409 (XXX) of 1976. See also United Nations E/CN. 5/477 of October, 1972, E/CN. 5/519, December 1974 and UNRISD/ 76/C. 45 GE 76—12446, December, 1971.
3. J.M. Keynes, *The General Theory af Employment, Interest and Money* London 1936, Chap. 24.
   It appears Keynes wanted to present a general theory; in actuality, however, it turned out to be a special theory of depression. And like Marx, who was considered by Keynes as classical, Keynes himself has now become both classical and incomplete. It is understood that Keynes wanted to completely revise his work, but ill health prevented him from attempting a re-writing.
   See also J.R. Hicks, particularly his articles on 'General Theory' in *Readings in Income Distribution*, AEA (London 1966).
   V. Shanmuga Sundaram, "Towards a General Theory of Social and Economic Planning" (mimeo): Paper presented to State Planning Commission Seminar on Social and Economic Planning for Integrated Rural Development, Madras, October, 1976.
4. S. Kuznets, Economic Growth and Income Inequality, *American Economic Review* 45 (March), 1955, 1—28.
   See also Quantitative Aspects of Economic Growth of Nations; III, Distribution of Income by Size. Economic development and cultural change II (January), 1963, 1—80.
   See also the Gap: Concepts, Measurement and trends. In *the Gap between Rich and Poor Nations*, ed. G. Ranis, London: Macmillan. 1972.
   J.G. Williamson, Regional inequality and the process of national development: A Description of the patterns, *Economic Development and Cultural Change*, Vol. 23, 4. part 2, 1965, pp. 3-45.
   I. Adelman, & C. T. Moris, *Economic Growth and Social Equity in Developing Countries*, Stanford, 1973.
   H. Chenery & M.S. Ahluwalia *et al.*, *Redistribution with Growth*, Oxford University Press, 1974.
5. Writers have acknowledged the increased data availability in many of these countries, but have expressed concern about their lack of statistical quality. According to Ahluwalia, the data sources used in

many studies in developing countries relating to the estimation of income distribution and other development indicators may be "the best available, but are simply not good enough". See Montek. S. Ahluwalia, Income inequality : Some dimensions of the problem, in Chenery *et al.*, *Redistribution with Growth*. Oxford University Press, 1974.

6. Brookfield remarks : "The dimension of geographical space is merely one in which unequal advantage is most obvious".
   H. Brookfield, *Interdependent Development*, Methuen, 1975, page 123.
7. This term connotes inequalities of socio-economic organization including political groups, classes or economic sectors and institutions.
8. While an analysis of areal differentiation and spatial relationships does constitute a necessary dimension in the study of underdevelopment, it is not a sufficient condition for a total understanding of underdevelopment. Exclusive concentration on this dimension to the virtual exclusion of other dimensions would mean concentrating only on the shell and not on substance. Spatial solutions could be devised by reorganizing the spatial division of labour, by restructuring relative distances, by improving accessibility or reforming territorial boundaries. But the spatial solutions are insufficient to tackle the problems of inequalities. "The spatial structures which men create—their cities and towns, transport routes and terminals—do not have an existence independent of the societies which they reflect". Coates et al., *Geography and Inequality*, Oxford University Press, 1977, p. 253.
9. Coates, *et al.*, Ibid., p. 5.
10. Hollis Chenery *et al.*, *Redistribution with Growth*, Oxford University Press, 1974.
    Irma Adelman & Cynthia Taft Morris, *Economic Growth and Social Equity in Developing Countries*, Stanford University Press, Stanford, California, 1973.
11. The selection of these countries for the purpose of this study has no significance other than the fact that the facts relating to their regional economic development are somewhat more familiar to this author, who has not only visited these countries and collected a lot of information about them, but also has had the benefit of discussions with many research scholars in these countries.

CHAPTER 2

THE INEQUALITY PROBLEMS IN THE ASIAN COUNTRIES : An Overview

UNEVEN development seems to be a pervasive feature of all the developing countries. This unevenness in many Asian countries is reflected on the one hand in increasing income inequalities, unemployment, poverty and the lack of basic human needs and on the other in spatial terms such as rural urban disparities and spatial underdevelopment, that is the emergence of contiguous areas of backwardness possessing a syndrome of collective poverty. Various factors have led to this state of affairs. We will be going into these reasons in depth in this book. In this chapter, it is proposed to provide a general introduction to the inequality problem in the Asian countries. This is followed by a selective review of theories bearing on this problem and an in-depth analysis of the regional inequalities problem in five countries of South-east Asia namely, India, Indonesia, Malaysia, the Philippines and Thailand. These five countries, for our present analytical study, constitute the study region. For the purpose of our comparative analysis, a few more Asian countries have also been brought into this discussion.

The percentage annual growth rates of population, gross domestic product (GDP) agricultural output, GDP per capita and agricultural output per capita of these countries during 1950-1973 may be seen in Table 2.1. This table also shows, for the sake of comparison, the growth rates for a few other countries in the East and South Asian regions.

Taking the rate of growth of gross domestic product as a measure of development, it may be seen from Table 2.1 that among the countries, India has grown at a slower p ace than the other four countries in the study region. Among

TABLE 2.1. **Percentage annual growth rates of population GDP, GDP per capita, agricultural output and agricultural output per head (1950-73)**

Country	Population (Growth rate in percentage)	GDP (Growth rate in percentage)	GDP per capita	Agri-cultural output (Growth rate in percentage)	Agri-cultural output per capita
1	2	3	4	5	6
India	2.0	3.9	1.8	2.8	0.8
Indonesia	2.1	5.4	2.2	3.2	1.1
Malaysia	2.9	5.1	2.1	4.0	1.1
Philippines	3.0	5.8	2.8	3.4	0.4
Thailand	3.0	6.7	3.5	4.3	1.3
South Korea	2.4	7.7	5.1	4,7	2.3
Pakistan	3.2[a]	6.0	2.8	4 8[a]	1.6[a]
Burma	2.0	4.7	2.7	2.4	0.4
Sri Lanka	2.5[a]	4.1	1.7	2.3	—0.2
Nepal	1.8[a]	2.3	0.6	1.4[b]	—0.4[b]
Bangladesh	2.8[a]	2.0	−0.7	1.7[a]	—1.1[a]

Source : *IBRD*, World Tables 1976, Washington, 1976.

[a]1960-1973

[b]1961-1973

other countries, South Korea has shown a spectacular rate of growth, while Nepal and Bangladesh have growth rates which are below those of the other countries in the region.

Since population in the Asian countries has also grown significantly during this period, a comparison of growth rates of per capita output (which is more meaningful) has also been indicated in col. 4 of Table 2.1. This shows that Bangladesh is the only country where the output per head failed to rise. In all the other countries, it has increased somewhat spectacularly in the case of some and moderately in others. South Korea is an instance of spectacular increase (5.1 per cent per annum respectively). All the countries in the region are predominantly agricultural and all of them have been increasing their agricultural output. In terms of per capita agricultural output, however, there is decline in three countries (Bangladesh, Nepal and Sri Lanka), slow growth in Burma and the Philippines

(0.4%), moderate growth in India (0.8%), Indonesia (1.1%), Malaysia (1.1%) and fairly high growth in South Korea (2.3%).

While the development trends and experiences of the individual countries in the Asian region have varied greatly, there is no doubt that there has been sustained growth of domestic production in these countries. Although international comparisons have their serious limitations, analysts have not been wanting to point to a fact that the rates of growth experienced by a number of Asian countries have exceeded, sometimes by large margins, the rates of growth experienced by the new industrialised countries during the period of their accelerated expansion, approximately a century ago.[1]

Notwithstanding the considerable improvements in aggregate growth performances revealed in this kind of global analysis, there is considerable evidence—albeit indirect and fragmentary—to show that developments in South and South-east Asian countries have been highly uneven. Some sectors, regions and classes seem to have benefited considerably more than the others, and there appears to exist great inter-regional inequality in resource endowments, in the organisation of production and in social consumption and welfare services, all of which are ultimately reflected in the distribution of income among the people. The lack of reliable income statistics at sub-national levels in these countries, on a comparative time-series basis, prevents us from presenting a comparative picture of the magnitude and pace of development at intra-national scales. This paucity of data has stood in the way of an adequate multi-national comparative analysis of the problem of inter-regional inequalities in the developing countries. In Williamson's cross section analysis of inter-regional income disparities for 24 countries, the developing countries of Asia are represented only by two countries namely, India and the Philippines, which have coefficients of inter-regional disparity of 0.275 and 0.556 respectively.[2] Yet, the conclusion is significant namely, that the poor but developing countries are characterised by increasing regional disparities, while in the more developed ones, the trends are perhaps towards a greater degree of inter-regional equalisation.

The marked presence and persistence of regional inequalities in the developing countries of South and South-east Asia has received some attention in a number of recent studies. The generally pronounced dual economy characteristics of the developing countries have been commented upon in a number of studies. They are often interpreted as a dependency problem. It has been pointed out that development has tended to be concentrated in one, occasionally more, dominant metropolitan centre (core area or enclave) with the hinterland (or periphery) both lagging behind and also exposed to the changing ways dictated by the core area. While the basic pattern owes to colonialism, its continued persistence is attributed to the role of overseas capitalism, internal colonialism and the dominance of the primate city which are all the characteristics of economic structures in the developing countries. Stohr[3] has pointed out that industrial activities are much more concentrated in and around major cities in the newly developing countries, such activities usually extending wave-like over only a short distance from the major metropolitan centre and rarely residing in secondary or lower-ranking cities. Logan's studies[4] on Nigeria have revealed that the development of manufacturing in the country has led to some import replacement, but has not effected any restructuring of the space economy. The striking disparity between urban and rural incomes in the developing countries of Asia (which have also been widening over time) has been identified by some authors as the basic underlying cause of poverty in these countries.[5] In the Philippines, Thailand, Malaysia, India, Bangladesh and Sri Lanka (including the plantation sector), average incomes in rural areas are a half or less of those in urban areas. In Indonesia and Pakistan, rural incomes are about a third lower than urban. Only in South Korea, the rural urban income differentials are rather narrow. Table 2.2 shows the rural urban differentials in income in selected countries and also the income share of the highest and lowest quintiles of population.

From the available evidence on the trends in income distribution in the rural areas, it may be seen that of the eight countries for which we have data (Table 2.3), in none of the five countries of our study region, the distribution of income as measured by the Gini coefficient has improved. On the other

TABLE 2.2 : **Urban-rural income disparities in some Asian countries**

Country	Income						Rural/ Urban Ratio	Income Share of	
	Year	Currency Unit	National Mean	Urban Mean	Rural Mean			Highest Quintile %	Lowest Quintile %
Philippines	1971	Pesos	3,736.3	5,865.6	2,817.6		0.48	55.0	4.0
Thailand	1968-69	Baht	13,260	24,277	10,403		0.43	51.6	6.9
Malaysia*	1970	M $	275	441	204		0.46	56.2	4.1
India	1964-65	Rs	1,580	2,700	1,360		0.50	48.1	7.8
Sri Lanka	1969-70	Rs	289.1	452.3	256.6		0.57	44.7	8.2
Taiwan	1971	NT $	48,700	51,600**	40,900		0.79	37.3	8.7
Japan	1971	10,000	143.4	146.4**	123.0***		0.84	43.1	5.5
Indonesia†	1969-70	Rupiahs	1,404	1,877	1,317		0.70	—	—
Pakistan††	1970-71	Rs	235	317	209		0.66	—	—

Compiled from different sources by Fuchen Lo and Kamal Salih in their paper "Regional Policy and Rural Urban Transformation in Asia", Ibid Table 8.

Sources : Derived from Bureau of Census and Statistics, Republic of the Philippines. Family Income and Expenditure Survey, 1971; Oey Astra, Meesook, Income Distribution Statistics for Thailand, Malaysia and Indonesia, August 1974, (mimeo); Post Enumeration Survey of the 1970 Census, Malaysia; Pranab K. Bardhan, IBRD. The Pattern of Income Distribution in India : A Review, June 1973 (mimeo); Socio-Economic Survey of Ceylon, 1969-1970, Preliminary Report, Table 30; Survey of Family income and Expenditure, Taiwan, 1971; Employment Status Survey, Japan, 1971. Table 63. In Harry T. Oshima, "Perspective in Income Distribution, Research "Income Distribution, Employment and Economic Development in Southeast and East Asia. Papers and proceedings of the Seminar sponsored jointly by the Japan Economic Research Center and the Council of Asian Manpower Studies, December 16-20, 1974 (Tokyo and Manila; July 1975), Charts 1-7, pp. 26-32. Also from Wouter von Ginneken, *Rural and Urban Income Inequalities in Indonesia*, Mexico, Pakistan, Tanzania and Tunisia, A WEP Study, (Geneva : International Labour Office, 1976), Tables 2 & 9, pp. 12 & 26.

*Figures for Malaysia are in monthly income.

**Non-farmers for Taiwan and non-agricultural for Japan.

***Farmers for Taiwan and agricultural for Japan.

†Figures for Indonesia are household consumption expenditures in rupiahs per month per household member.

††Figures for Pakistan are household income per month.

hand, the degree of inequality seems to have worsened over the years. The degree of inequality in South Korea appears to have remained unchanged since 1966. Bangladesh and Pakistan show anomalous pictures and the improvements indicated by their Gini coefficients have to be interpreted with caution. In the case of Bangladesh, the period covered by the two observations are very short and there are numerous indications that since the middle 1960s, the distribution of income in rural areas has become far worse. As regards Pakistan, the Gini coefficients suggest that rural inequality declined between 1963 and 1968 and then remained stable. Other evidence, in contrast, shows that inequality (as measured by the proportion of the rural population in poverty) probably increased after 1968 or 1969

TABLE 2.3 : **The distribution of household income in the rural areas of Asia**

(*Gini coefficient*)

Country	Year	Coefficient
Bangladesh	1963-64	0.35
	1966-67	0.33
India	1953-57	0.31
	1960	0.45
	1964-65	0.37
	1967-68	0.48
Malaysia	1957-58	0.35
	1970	0.46
Pakistan	1963-64	0.36
	1966-67	0.33
	1968-69	0.30
	1969-70	0.30
	1970-71	0.30
Philippines	1961	0.41
	1965	0.43
	1971	0.47
South Korea	1966	0.31
	1970	0.31
	1971	0.31
Sri Lanka	1969-70	0.35
	1973	0.37
Thailand	1962-63	0.44
	1970	0.45

Source : Shail Jain, *Size Distribution of Income : A Compilation of Data*, World Bank, Washington, 1975.

and that by 1971-72 inequality had once again become as great as in 1963.

From whatever angle we approach our analysis, the results support the contention that the regional problem is quite acute and extensive in the developing countries of South and South-east Asia. The manifestation of this uneven development is sometimes described as "the crisis in the rural areas of Asia". Keith Griffin has in a recent penetrating study analysed the various explanations for this rural crisis and has concluded that it has little to do with growth rates, either national or sectoral; it is primarily a *distributive crisis*. rooted in the problem of growing inequality in the distribution of income levels within the countries.[6]

Some of the manifestations of inequality in the developing countries are : Poverty, inequality in the distribution of assets, particularly land, unemployment and rural-urban imbalances. Ahluwalia, Carter and Chenery have recently analysed growth and poverty in developing countries. The Table 2.4

TABLE 2.4 : **Variations in poverty levels**

Country	GNP per capita for 1975 (in 1970 US)	GNP rates 1960-1975	Population growth rate 1960-75	% of population in poverty 1975	Share of lowest 40% popn. in GNP 1975
India	300	3.6	2.3	46	17.0
Indonesia	280	5.2	2.1	59	16.1
Malaysia	1006	6.7	2.8	12	11.1
Philippines	469	5.6	3.0	33	11.6
Thailand	584	7.5	3.0	32	11.5
South Korea	797	9.3	2.1	8	16.9
Pakistan	299	5.6	3.2	43	16.5
Burma	237	3.2	2.2	65	15.7
Sri Lanka	471	4.2	2.4	14	19.3
Nepal	—	—	—	—	—
Bangladesh	200	2.4	2.7	64	20.1

Source : Tables 1 and 2 in Montek S. Ahluwalia, Nicholas G. Carter and Hollis B. Chenery: Growth and Poverty in Developing Countries, World Bank, Staff Working Paper No. 309 (Revised) May 1979. Using Kravis adjustment factors.

abstracted from their work reveals the variations in poverty levels.

The poverty levels revealed by the above table are closely associated with extreme inequality in the distribution of assets, particularly land. According to Gunnar Myrdal, "particularly in the South Asian rural setting, inequality is mainly a question of land ownership with which are associated leisure, enjoyment of status and authority".[7] This has inevitable repercussions on the process of economic development and is reflected in the growing numbers of landless labourers in Asian countries. The nexus between property ownership and economic power is also the interrelation between certain aspects of private property law and public investment programmes in such diverse areas as irrigation, road building, land development or cooperative movement. The development pendulum has always swung in favour of the privileged few in these countries. This seems to be the experience in most developing countries.

The World Bank analysis has clearly revealed that the distributional pattern of past growth has left the poorest groups largely outside the sphere of economic expansion and material improvements. The core of the problem of poverty in these countries is the lag in the growth of income of the poorest 40 per cent of the population. The favourable position of South Korea among this group of countries may be noted. It has been commented that this achievement is related to the pursuance of policies directly affecting income distribution. In any case, there is substantial agreement today among development economists that the benefits of growth accruing to the poor can be increased through distributional policies.

Another important dimension of inequality in the developing countries is the widespread prevalence of unemployment and underemployment in the Asian countries. The varying concepts of employment and unemployment adopted in the various countries and the diverse methodologies adopted for their measurement preclude us from quoting some authentic figures to bring out their magnitude in the different countries.

Nevertheless we may note that the presence of some sectoral incapacities or constraints has been identified in these countries. The insufficient domestic demand for goods and hence the small domestic market size as well as various barriers to penetration of foreign markets are usually cited as some constraints. Both the manufacturing as well as the agriculture sectors in these countries suffer from such constraints, which seriously limit the room for labour absorption in these sectors. The surplus labour in these countries, therefore, being unable to find opportunities in the productive sectors of the economy, have drifted into low income, low productive service occupa-ʼons, thus swelling the informal sector both in the rural and ʼban areas.

)hima's study[8] on the pattern of labour absorption in East South-east Asian countries among three broad sectoral egories namely, agriculture, manufacturing and services, has ʼarly brought out the low labour absorptive capacity of the gricultural sector in these countries. This low labour absorptive capacity in agriculture may be due to several reasons such as for example, the generally low income elasticity of demand for agricultural products; the bias of government policies towards urban-industrial development in the 1950s and 1960s; the export orientation in agricultural products resulting in a somewhat disembodied technical progress (such as better seeds, fertilizers) in cash crops (for example in Malaysia and Sri Lanka), the capture of productivity increases by large scale farmers using labour saving devices such as those related to production, etc. Ohima's analysis also reveals that in the manufacturing sector, the growth rates in these countries are not high, excepting in South Korea. Nor was labour absorption any significant in the service sector naturally so, as the expansion in this sector is related to the service requirements of manufacturing and modernisation of agriculture. As a result, there is a high degree of unemployment and underemployment in these countries. The spatial reflection of these imbalances may be seen in the phenomenon of primate urban growth. Thus, 55 per cent of Thailand's urban population is concentrated in Bangkok; 36 per cent of South Korea's urban population in Seoul, 21.4 per cent of Indonesia's urban population in Jakarta and 21 per cent of Malaysia's urban population in Kuala Lumpur. A corollary of this pattern of urban growth is the inadequate intermediate urbanisation in these countries which has accentuated spatial inequalities. In the case of India, Sri

Lanka, Thailand, Indonesia and West Malaysia, a large majority of their population is still rural.

Patterns of inequality

In multi-national comparative studies, our analysis and understanding of problems get greatly improved by identifying certain similarities in development characteristics or behaviour and classifying countries on these lines. Such categorisatior although broad and descriptive, greatly facilitates poli oriented discussions. As early as 1955, Hoselitz used historical pattern of growth as a criterion for classifying s countries.[9]

Fuchen Lo and Kamal Salih[10] have recently assembl data relating to GDP per capita and the role of primary export in GDP, and have also worked out Gini income coefficient as well as Gini coefficient of land concentration for selected Asian countries in order to identify regional typologies, if any, in respect of income and other inequalities in these countries.

Their studies have shown that the Gini income coefficient is evidently the highest among the South-east Asian group (Malaysia, Thailand, the Philippines and Indonesia) as compared with the countries in East Asia (Korea, Taiwan and Japan). The South Asian countries (Sri Lanka, Bangladesh Pakistan and India) however, tend to exhibit a higher variability among them with Bangladesh showing greater equality (comparable to Japan, but of course at lower level of GDP per capita) and India at the other extreme. Taiwan and Korea have considerably improved their Gini coefficient, clearly showing that reduction of inequality in income distribution is accompanied by rising per capita incomes.

When income inequality is related to the percentage of primary export in GDP of each country, (which is a rough indicator of its economic dependency), the extreme position of Malaysia and the intermediate position of Sri Lanka and India are evident. In regard to coefficient of land concentration, it is clear that the South Asian countries (Sri Lanka, Bangladesh, Pakistan and India) have a generally higher degree of inequality. In addition, it may also be seen that in every case, except Thailand and in a marginal way the Philippines (where some land reform has been undertaken), the land concentration ratio has either stagnated at a high level or in fact deteriorated. Notable among the latter are Bangladesh and Malaysia, each of which had increased ten points, and to a lower extent India, in more or less the same number of years.[11]

It would be reasonable to assume that the patterns of inequality existing in the Asian countries are to be related partly to the external dependency of the economy and partly to the internal social structure. The degree of impact of these two factors in influencing national patterns of uneven development however would differ from country to country depending on the degree of openness of national economies, the nature of the national and regional policies being pursued by them, the extent of foreign ownership of assets, the volume of foreign investment and aid and the flows of economic surplus etc. The analysis of Fu-chen Lo and Kamal Salih has shown that among the Asian countries, the dependence on the industrialised countries is the most severe in the primary producing countries of South-east Asia (Malaysia, Philippines, Thailand and Indonesia) than in any other regional group. Again among the South-east Asian countries, Malaysia and Singapore show a pattern of openness comparable with some countries in east Asia such as Japan, Taiwan and Korea.

Based on the structural characteristics and transformation processes operating, the Asian economies have been classified into four models, namely, (i) Resource rich open dependent economy (represented by many South-east Asian countries); (ii) low per capita resource semi-open economy (represented by most South Asian countries) ; (iii) centrally planned closed economy (represented by China) ; and (iv) industrialised semi-open economy (Japan before 1960s and Korea).

Their structural characteristics and transformation processes are indicated in Table 2.4 on the next page.

Spatial inequalities have now undoubtedly become an important policy issue in the developing countries. During the last two decades, in particular, many countries in the Asian region have evolved broad regional policies, a number of which are oriented towards reducing the spatial disparities of living levels at the national level. Such policies have emerged both for objective and subjective reasons. Objectively because, in most of these countries, de-colonialisation had led

TABLE 2.4 : **A classification of Asian economies**

Model	Structural characteristics	Transformation process
1. Resource rich open dependent economies (e.g. SE Asian countries)	(a) Relatively rich resource endowments (b) Relatively open domestic market (c) Relatively low technology	(a) Foreign investment in resource and commercial crop sectors (b) Resource-based primary product export (c) Foreign investment in modern industry (d) Import substitution-modern industrialisation (e) Rural development expenditures (f) Kinship based economic remittances and some return migration (g) Excessive rural-urban migration (h) Tenant-landlord relationships
2. Low per capita resource semi-open economy (e.g. South Asian countries)	(a) Low per capita resource availability (b) Remote rural area with low urbanisation (c) Relatively low technology	(a) Limited commercial crop export potential (b) Considerable foreign aid (c) Import substitution industry with modern technology and limited manufacturing export capacity (d) Tenant landlord relationships (e) Low urban accessibility and moderate rural-urban migration. (f) Socio-institutional rigidity (social immobility).
3. Centrally planned closed economy (e.g. China)	(a) Food deficit with high population density (b) Closed economic and political	I. *First Phase Transformation* (a) Redistribution of assets (land, capital etc.)

Model	Structural characteristics	Transformation processes
	system (c) Indigenous technology	(b) Controlled production and distribution system (c) Controlled rural-urban migration. II. *Second Phase Transformation* (d) Minimise external dependency (e) Self-reliant regional development (f) Rural-based urbanisation (agrofeed and agro processing industrialisation)
4. Industrialised semi-open economy (e.g. Japan before 1960's Korea)	(a) Indigenous modern technology (b) Poor natural resource endowment with high population density. (c) Semi-open economy with protected domestic market.	(a) From import-subsituation to export-oriented indusrialisation. (b) High labour absorptive capacity and urbanization (c) Post-war land reform (d) Complementary linkages between large and small scale industries (e) Rural infrastructure formation

Sourcc : Fuchen Lo & Kamal Salih, *Regional Policy and Rural-urban Transformation in Asia*, Ibid.

to attempts towards the integration of sub-national areas, seeking to break the enclave structure of the economy and bring about the transmission of development over much wider areas. This is gradually leading to a change in the spatial structure of underdevelopment. Subjectively, spatial disparities have become important policy issues because with the increasing scale of interaction and information exchange, there has been an increasing perception and articulation by local and regional communities of the spatial inequalities and the need for removing these disparities as quickly as possible.

This point (which is generally valid for all countries in the Asian region) needs some elaboration. It relates to a changing orientation and structure of development planning in these countries. In the past, the main objective in development planning was primarily on achieving high growth rates of total GNP (the so called top-down approach). Such an approach led to an aggregated type of planning in which the sub-national and sub-regional level issues were neglected or given only secondary importance. Now there is an upsurge for the decentralisation of the planning process to the sub-national levels. Regional and local development issues are claiming greater and greater attention by national governments. Greater decentralisation (that is, spatial disaggregation with devolution of power and authority to lower authorities) of national development planning with creation of multi-level planning structures informed by varying goal concepts related to different levels is being attempted in these countries. There is a growing preference by governments for adopting small geographical, but viable local areas as the basic units of planning and implementation and then build upwards. New development strategies, variously termed—"planning from below", "planning for growth with social justice", "micro-level planning for full employment", "people-oriented strategy of development" and so on are being vigorously debated in the developing countries.

TABLE 2.6 : **Structure and orientation of development planning—Global trends**

		Public Sector Investments	GDP Growth	Redistribution with Growth	BMN Approach, Employment and growth focussed on small areas
		ORIENTATION			
Centralised Planning	STRUCTURE				
Deconcentration					
Decentralisation					
Local self-reliance					

The tendencies discernible in the orientation and structure of development planning during the last two decades may be generalised as in the Table 2.6.[12]

In many countries, specific policy interventions have taken place for promoting the development of regions which lag behind the national average or for controlling development in areas of extremely rapid economic growth, which are experiencing severe problems of agglomeration and congestion. Besides policies aiming at spatial strategies, the countries are also adopting various kinds of income distributional policies to increase the benefits of growth accruing to the poor. They seek to

(a) increase linkages of the poor to the faster growing segments of the economy so as to increase the flow of indirect benefits; and

(b) provide much greater direct support to productive activities upon which the poor are heavily dependent and which have a potential for efficient expansion.

The present work is confined to the five countries in South-east Asia and is designed to analyse the nature and origin of underdevelopment in these countries, the policies that have been devised and the impact of such policies.

NOTES AND REFERENCES

1. See W. Arthur Lewis, *Growth and Fluctuations (1870-1913)*, London George Allen and Unwin, 1978. Also Keith Greffin, Growth and improvement in the Rural Areas of Asia (Mimeo) Paper contributed to Round Table on Adaptation of Administration to Rural Development organised by Govt. of India and ESCAP, August 16-18, 1978, New Delhi.
2. The coefficient of inter-regional disparity (Vw) measures the deviation of region 1 per capita income from the national average and is weighted by the relative magnitude of rural regional population; the higher the Vw, the greater is the interregional income disparity. In Williamson's study of 24 countries, the lowest value is 0.058 for Australia and the highest value is 0.700 for Brazil.
3. W.B. Stohr, Inter-urban systems and regional economic development. Resource Paper No. 26, Commission on College Geography, Association of American Geographers, Washington, 1974.
4. M.I. Logan, The development process in the less developed countries,

Australian Geographer 12, (1972).

5. See for example, Michael Lipton, *Why Poor People Stay Poor: Urban Bias in World Development*, London, Temple Smith 1977.
6. Keith Griffin, Growth and impoverishment in the rural areas of Asia, Paper presented to the Round Table on Adaptation of Administration to Rural Development organised by Govt. of India and ESCAP, 16-1 August 1978. The various alternative hypotheses, tested by Griffi include: Cyclical instability of food supplies, demographic phenomen that is density of population and growth performance of the agricultural sector.
7. G. Myrdal *Asian Drama*, New York; The Twentieth Century Fund 1968, Vol. I, p. 569.
8. H.T. Ohima, Labour absorption in East and South-east Asia : Summary, Perspective and Prospects, *The Philippines Economic Journal*, volume 15, Nos. 1 and 2 (1975).
9. His examples are as follows: (i) Expansionist, dominant, autonomous—the United States from 1830-90, (ii) expansionist, dominant, induced—Soviet Union, 1928 to present (iii) expansionist, satellite, induced—Australia to 1914 (iv) expansionist, satellitic, induced—Manchuria under Japanese control; (v) intrinsic, dominant, induced—Japan or Turkey since 1922; (vi) intrinsic satellitic, induced, the so-called "people's democracies" in Eastern Europe. (viii) intrinsic, dominant, autonomous-19th century, France or Germany, See Best F. Hoselitz, Patterns of economic growth *Canadian Journal of Economic and Political Science*, Vol. XXI, 1955.
10. Fuchen Lo and Kamal Salih *et al.* "Regional Policy and Rural Urban Transformation in Asia." U.N. Centre for Regional Development, Nagoya (Japan), November, 1977—Policy Research Programme Working Paper (mimeo).
11. Fuchen Lo, Kamal Salih *et al.* op. cit. the authors consider that no clear regional typology has emerged in this regard.
12. W.H. de Man and A.D.J. Sehaap, A Model for incorporating the element of space and some aspects of the basic human needs approach in regional development, *ITC Journal*, Vol. I, 1979, Enschede, Netherlands.

CHAPTER 3

SPATIAL INEQUALITIES AND BACKWARD AREA DEVELOPMENT: CONCEPTS, POLICIES AND STRATEGIES

THE PROBLEM of inequalities in the process of development has been the subject of intensive research since the 1950s and quite an extensive, but heterogeneous, literature has accumulated over the years. Notwithstanding this impressive documentation, it must be admitted that a full-fledged spatial theory of development or underdevelopment (that is the regional development of backward areas) is yet to come. However, there are a good many empirical generalisations, hypotheses and partial theories to explain specific phenomena of regional growth and development as well as underdevelopment. It may be assumed that the building blocks for a theory of development of backward areas already exist in the development literature.

The object of this chapter is broadly to review the various theories and concepts bearing on the pattern of spatial inequalities. This is being attempted with an eye on the policy aspect. All sound policies and programmes must stem from a rigorous conceptual framework and analysis. Otherwise, they will tend to be ad hoc, inadequate and ill-conceived. Many developing countries of the world are now engaged in experimenting with new policy instruments aiming at interregional equalisation of development benefits. In this context, the need for pinning down the policy framework to an appropriate conceptual base cannot be over-emphasised.

The problem that is sought to be handled here is an exceedingly broad and involved one. Theoretical literature in economics and other social disciplines dealing with the process of development in general, and with its geographical dimensions in particular, has accumulated on a vast scale since the 1950s. Several empirical studies have been made and theories, hypotheses and models have been proposed to explain how development occurs and how it spreads in both economic and geographic spaces. Theories have also been advanced to explain the spatial structure of underdevelopment in the developing countries. Some "stage of growth" models have also been suggested. But no single theory has proven to be without flaw; for in real-life situations, there generally appear a number of variables, which a theory or a model has not taken into account. Consequently, those theories and models have to be revised or abandoned or put only to limited use. Nevertheless, it must be admitted that our understanding of the development process has increased, as we have identified an increasing number of interrelationships among development variables. The increasing concern over uneven internal development in Third World countries has also led to some fresh thinking on new paradigms of development. Thus, poverty-oriented strategies, employment-oriented strategies and planning from below through more self-reliant and participatory approaches to development planning have emerged, While seeking a way through this jungle of confusing literature, one is likely to be lost, unless one keeps an over-view and is discriminate with reference to one's ultimate goals. We therefore commence our review with some delimitations and structuring. Otherwise, the argument is bound to become diffused and uncoordinated. Broadly, the numerous theoretical works relating to the problem of inequalities and regional development may be divided into the following groups :

1. Theories that seek to explain how development spreads in economic or functional space.
2. Constructions based on the concept of growth through "geographic clusters" (poles of growth).
3. Theories that regard development as a spatial diffusion process.
4. Economic base type theories and related "stages of growth" models designed to explain how a region grows.
5. Hypotheses concerning convergence and divergence of regional inequalities.

6. Theories of spatial structure based on Marxian analysis.
7. Theories based on the sociological contours of the problem: for example redistribution with growth, employment and poverty-oriented strategies.
8. The bottom-up paradigm of development.
9. Synthetic models of economic and social development.

Based on the above classification, the various theoretical streams in the literature bearing on the problem of regional inequalities and development are proposed to be identified and reviewed. More specifically, an attempt will be made to capture the central ideas running through many works, to juxtapose them, to disentangle some cobwebs and to highlight those insights which are fundamental to our discussion and of relevance to a theory of spatial inequalities and backward area development.

THEORIES THAT SEEK TO EXPLAIN HOW DEVELOPMENT SPREADS IN ECONOMIC OR FUNCTIONAL SPACE

The classical economists did not very much concern themselves with variations in development over geographical space, although it must be mentioned that some of their discussions prompted questions of a spatial order. Schumpeter[1] gave the lead to development thought by identifying it with innovations that is, the emergence of qualitatively new phenomena within the economy. Later writers reflected on how such innovations can spread. It was since the 1950s that development theory concerned itself more specifically with the aspects of concentration and spread of development.

Nurkse and others argued that in the context of the presence of a circular constellation of forces tending to act and react upon one another in such a way as to keep a poor country in a state of poverty (what has been termed the vicious circle of poverty), it was necessary that investment should be *diversified* over a broad range of industries. A wide range of products in different industries will support each other. "In this way, the market difficulty, and the drag it imposes in individual incentives to invest, is removed or at any rate alleviated by means of a dynamic expansion of the market through investment carried out in a number of different industries.... Through the application of capital over a wide range of activities, the general level of economic efficiency is raised and the size of the market is enlarged".[2]

The concept of balanced growth was challenged on several grounds. Fleming[3] accused the theory of overemphasis on the demand side of the problem. Inadequate supplies of factors of production in developing countries—particularly capital and skilled labour—may make diversified and simultaneous investment projects unfeasible by sharply increasing the cost of production of all industries, as they would compete for the limited supply of factors. Other arguments against the balanced growth approach were: (a) that it implied a closed economy; (b) it would mean a large public sector; and (c) it would require resources of such an order of magnitude, which an underdeveloped country cannot possess.

In the late 1950s, unbalanced growth became the central theme of a series of theories propounded by Hirschman and Myrdal. Hirschman maintained that growth is communicated from the leading sectors of the economy to the others. He was, therefore, of the view that development strategies should concentrate on a relatively few sectors, such sectors being determined by their ability to induce backward and forward linkage effects. For Hirschman, growth is necessarily an *unbalanced process* and takes place through a "chain of disequilibria", the expansion of one industry creating external economies for the other. He emphasised interindustry linkages—forward and backward linkages—leading to the establishment of a growing point. In course of time, as the polarisation effects[4] become unfavourable, trickle down effects will start. The trickle down effect[5] is a process of development of backward areas through product and factor input demand pulls from the developed region and is considered as an evolutionary phenomenon in Hirschman's analysis. According to him in the long run, public investment will cease to be pulled so heavily into the developed areas, largely because of considerations of equity and national unity. Thus in the long run, regional differences will tend to disappear.

The trickle down mechanism has therefore an important place in Hirschman's concept of development. According to him, scarce resources cannot be invested everywhere at once, and certain projects within certain sectors must obviously be selected for their growth potential. Uniform development over a whole country will take place, inevitably but gradually, and will be brought about through trickling down effects. It is also presumed that the state will intervene to influence the growth of imbalances whenever the normal market mechanism proves inadequate.

Myrdal's theory[6] of circular and cumulative causation, published at about the same time, enshrines similar concepts to those of Hirschman, but developed independently. Myrdal concentrated directly on the problem of inequality and employed concepts of backwash and spread effects (compare the polarisation and trickling down effects of Hirschman). The backwash effects involve the workings of population migration, trade and capital movements. The spread effects operate mainly through trade via interregional complementarities and are reflected in the increased outlets for the hinterland's agricultural products and raw materials and the tendency for technical advance to diffuse from the growth centres. The spread effects will be stronger, the higher the level of economic development and the higher the rate of growth of the economy of a country. Moreover, the attractiveness of the growth centre may be weakened by increasing external technological diseconomies and high labour costs. Finally, the governments of the wealthier countries are likely to initiate policies directed towards greater regional equality.

The contributions of Hirschman and Myrdal brought the problem of inequalities as an important issue in development literature. Myrdal clearly stated that lasting interregional imbalance in the less developed countries constituted an impediment to development and a part of the "interlocking relations by which, in the cumulative process, poverty becomes its own cause". Although similarities are present in the approaches of Hirschman and Myrdal, there are also differences of emphasis. In particular, Hirschman was more optmistic about the long run equilibrating effects of development; he seemed to take for granted that strong forces will eventually turn the tide, once the polarisation effects had proceeded up to a point.

CONSTRUCTIONS BASED ON THE CONCEPT OF GROWTH THROUGH "GEOGRAPHIC CLUSTERS"

Another important theoretical contribution which sought to describe and explain the anatomy of economic development in abstract economic space and which was developed almost simultaneously was the concept of growth pole which was triggered off by the ideas expressed by Perroux.[7] Perroux's primary concern was on economic growth and with the interactions among industrial sectors, rather than spatial development processes. He considered economic space as a field of forces consisting of centres, poles or foci, from which centrifugal forces emanate and to which centripetal forces are attracted. Each centre, being a centre of attraction and repulsion, has its proper field which is set in the field of other centres.[8] Perroux observed that "growth does not appear everywhere at the same time; it manifests itself in points or poles growth, with variable intensity; it spreads by different channels and with variable terminal effects for the economy as a whole".[9] Like Schumpeter, Perroux stressed the importance of entrepreneurial innovations in the development process, which proceed by a succession of dynamic sectors or poles through time. Another key element in Perroux's development theory was the concept of dominance. The idea of propellent industries exerting a considerable influence on the activities of suppliers and clients is an important aspect of his theory.

The spatial implications of Perroux's theory were elaborated by his colleague Jacques Boudeville. He emphasised that a complex of activities agglomerated around a propulsive centre was a development pole and the spatial planning problem was one of urban growth strategy. Thus the emphasis was on a few dynamic sectoral clusters and upon urban industrial growth as the key to more generalised regional development.

To him a regional growth pole is "a set of expanding industries located in an urban area and including further development of economic activity throughout its zone of influence".[10] However, in a regional setting "it would be preferable to describe poles as a geographical agglomeration of activities rather than as a complex system of sectors different from the national matrix. In short, growth poles will appear as towns possessing a complex of propulsive industries".[11] Conceptualisation of various types of space, region and of the nature of geographical and functional interdependencies is, therefore, fundamental to Boudeville's theory. Economic space is classified into (i) homogeneous (in relation to properties and characteristics), (ii) plan space (with reference to operational decision-making to achieve specific regional goal). A region is distinguished from economic-space, being essentially a *continuous area located in a geographical space*. A polarised region, therefore, is described "as a heterogeneous continuous area localised in a geographical space, whose different parts are interdependent through mutual complementarity and interplay relations around a regional centre of gravity".[12]

Growth pole thus is both a functional and geographical entity and its hallmark is the existence of a "propulsive structure". According to Boudevill, propulsive industry has two characteristics : "(a) a direct and indirect dominating influence over all other activities; and (b) an oligopolistic concentration of industry, with price leadership and a keen sense of anticipation in the moves of its own sector as well as in the related branches".[13] In analysing these characteristics, he sought a synthesis of the theories of Perroux, Hirschman, Leontief Youngaon,[14] Rasmussen,[15] Myrdal and a number of French economists. The importance of oligopolistic concentration is highlighted to include the phenomena of self-financing, research and innovation which are facilitated by price domination.

The main significance of Boudeville's analysis, however, lies in its direct operational content. He propounded the growth pole approach as one of the regional operational models and although his generalisations stem out of application to the French situation, they nonetheless have significance for other countries, provided they are suitably modified for differences in socio-economic and other relevant variables.

Thus, the French School represented by Boudeville tried to use the growth pole concept in policy planning by incorporating certain analytical tools with defined theoretical foundations. This, while seeking to explain the growth phenomena, reduced the concept of growth poles to an *inter-industry cluster*.[16]

Several critical evaluations of the growth pole concept have appeared. The studies of Hansen,[17] Lasuen[18] and Darwent[19] may be particularly cited. Broadly, the points of criticism relate to three issues :

(1) the absence of a location element in the formulation, that is, where the functional growth poles are or will be localised in geographical space,
(2) the absence of adequate explanation for the growth and functioning of the growth pole, and
(3) how the transmission of growth in geographical space (that is, the temporal interaction between geographical and sectoral clusters) would take place.

In order to fill the first two gaps mentioned above, it has been suggested that the framework of the growth pole theory should be related to those of central place theories and industrial structure analysis.* Lasuen[20] noted that stability in the hierarchical ordering of cities is "obviously one manifestation of regional convergence in income levels and economic structure".

THEORIES THAT REGARDED DEVELOPMENT AS A SPATIAL DIFFUSION PROCESS

The ideas emerging from Schumpeter and Perroux that development springs from *innovation* and that innovation in a leading industry resulting in a pole induces further innovations among its dependent activities were pursued by several authors, who have advanced hypotheses explaining how the successive innovations originate, are diffused and adopted, finally leading to successive sets of growth poles. The contributions of Hagerstrand,[21] Brown,[22] Morril, Pitts[23] and Lasuen[24] may be specifically mentioned in this context. Following this idea, the growth pole concept has been expanded *to refer to a system of poles rather to a single pole.*[25] According to Berry, the literature bearing on the diffusion process has achieved a new "geographical paradigm of progress". It is also suggested that knowledge relating to the poles of growth, the channels of

*Conventional growth pole analysis seeks to explain big industry clusters; central place theory to understand service activity location; industrial structure analysis to relate national and regional growth patterns.

spread and the sequence of change can point to a means "whereby the spatio-temporal lag between innovative and backward regions can be narrowed by setting up new growth poles and new channels, and by changing the pace and sequence of spread".[26]

The role of the growth pole in the diffusion of innovations and benefits from development is normally seen in terms of two processes. Firstly, through the process by which impulses of economic change are transmitted in order from higher to lower centres in the urban hierarchy and secondly, by the process whereby the spatial incidence of economic growth is a functioning of distance from the central city. Ideally, the establishment of growth poles would permit innovations, investment funds, and other forms of information to diffuse from the central metropolis to the new growth centres in the provinces. Again, from the lower order centres, other growth impulses will be emitted to the surrounding area, thereby involving the rural and traditional *areas in the growth process. The first process is closely linked* with central place theory, which would mean that there would be overall movement of information and innovation from high order centres to lower order centres. The second process is important at the local level and concerns the ways in which the growth process affects the rurai areas. This process of diffusion is said to be a function of two sets of probabilities, the probability of contact and the probability of acceptance. There is still insufficient evidence however in support of the idea that these two sets of probability are a function of geographical distance.

Identifying the development process with that of the generation and adoption of innovations, Lasuen has argued[27] that the process of generation of innovations is really an international one. He has, therefore, urged the creation of multinational firms on a continental basis as a pre-condition for more rapid innovation diffusion in the developing countries. However, the role of multi-national corporations in developing countries is a very complex one. The surging nationalism, the emphasis on indigenous technology and the uncertain investment climate prevailing in the developing countries make Lasuen's proposition highly doubtful and unrealistic.

Closely related to the diffusion theories are some hypotheses on modernisation which have been advocated in recent years. The contributions of geographers, particularly those of Gould,[28] Riddell,[29] Edward Soja[30] and Tobin[31] may be mentioned in this context. These studies have attempted to show modernisation as a spatial diffusion process, assuming patterns of varying intensity and rate. The importance of the spatial dimension has been emphasised in these contributions as affording "the key focus toward understanding the phenomenon of modernization".[32]

Tormod Hermansen has sought to interrelate the fundamentals of growth pole theory, central place theory and innovation diffusion theory. Thus with various interpretations, modifications and extensions, proposed around the growth pole concept, its scope has been sought to be considerably broadened and reoriented so as also to include within it the normative issues of policy intervention and planning. It has been advocated as a strategy for speeding up the process of economic growth and interregional integration and equalisation. According to Berry, "alternative growth centres need to be established to do by fiat what *laissez faire* movement of the population prevents".[33] The growth pole strategy is seen by many as an intermediate solution between extreme concentration and extreme decentralisation. From the point of view of spatial disparities, it has been considered to have the inherent possibility of increasing the spatial imbalances for the same reasons that an economic development plan may cause sectoral imbalances, namely through the lack of adequate resources to develop simultaneously a sufficient number of poles to achieve the national spatial equilibrium.[34]

ECONOMIC BASE TYPE THEORIES AND RELATED "STAGES OF GROWTH" MODELS

The economic base concept, applied originally to the study of cities, was adopted in 1950s by North, Tiebout and Perloff to the study of regions. Drawing upon the fundamental theorem of the doctrine of comparative advantage in international trade, North, the original author of the economic base theory[35] argued that every region, however backward, had a comparative advantage in the production of some commodities. These commodities would constitute its export sectors in the sense that the large bulk of production was directed for export to other regions. It is argued that the export earnings accruing to the economic base should operate through the regional income multiplier, leading to the development of the area as a whole.[36] In a further elaboration of the concept based on empirical generalisation relating to the historical development of the Pacific North West region of the U.S.A., a sequence of five stages of growth of the regional economy—from a stage of subsistence economy through the specialisation in one or more of primary goods for export, and then a growing expansion and diversification of the export base by development of secondary activities to the stage of widely diversified economy including highly specialised services—was postulated.[37]

The above sequence of growth was also supposed to be applicable to other capitalist countries characterised by small population pressure, particularly countries with frontier regions, possessing clear comparative advantages owing to their superior endowment in natural resources; but this assumption has been questioned.[38] Whether such a strategy of development will succeed in a particular case will depend very much on the specific set of circumstances attendant to that particular case.

Watkins has developed a descriptive model called staple theory based on early Canadian economic growth. It shows how a fast-growing export based on local specialised natural resources (the staple) can induce an accelerated rate of investment and an inflow of labour and capital resources into the staple sector, as well as providing opportunities and inducements for industries in other sectors to meet these investment needs, providing infrastructural services and meet the growth in demand for consumption goods arising from the higher personal incomes generated in the staple sector.[39]

As regional economies are much more open compared to the national economy, there are also possibilities for considerable leakages of resources from the local area. Thus, whether or not the export sector of a backward area can serve as an effective basis of development of the area depends very much on the net balance of trade of the area. Furthermore, in the context of the economic and institutional structures existing in backward areas in developing countries, there is no guarantee that even a surplus balance of trade would necessarily lead to the development of a backward area. This is because of the presence of forces even within the peripheral area such as, developed urban enclaves and so on which tend to siphon off surplus from the backward areas. This point has been elaborated in a later context in our review. Viewed in this light, it would appear that the economic base strategy for backward area development under the existing system would only serve to consolidate the mechanism of surplus drain.

In order to overcome the weakness of the economic base strategy, variants of the strategy are sometimes suggested. One of the suggestions is to develop projects of comparative advantage in a backward area for export to other regions along with a support price policy so as to ensure that the surplus does not leak out along with export. But even such a policy may be only partially successful. Price support for instance can be provided only if a product is exclusively produced in the area in question. If this is not the case, such price support has to be extended throughout the national market. Therefore, a price support policy cannot be executed locally for a single backward area only. Even assuming that such a policy is successfully implemented, this alone will not ensure retention and investment of the resultant surplus. This is because in order that a policy for the developing backward area is successful, not only should surplus resource for investment be available but at the same time an internal demand should be generated. And since the export—with price support strategy—can do little to generate such local demand, it would do no more than link up the export sector with the urban industrial enclave, without altering in any way the backwardness of the area as a whole.

Other variants of economic base strategy tried in developing countries are the establishment of industrial estates in backward areas or the location of large industrial projects. The experience in both these instances clearly indicates that they do not generate the kind of demand impulses which would take care of the double development constraint namely inadequate supply of crucial inputs and inadequate local demand.

From Rostow's[40] original publication on the stages of

economic growth, a number of theoretical formulations are available, which are based on the idea of some kind of evolutionary sequence of economic transformation of countries. Rostow put forward the idea that all societies either have passed or will have to pass through the five stages namely, (i) The stage of traditional society; (ii) that of the preconditions of development; (iii) that of take-off; (iv) that of maturity; and, finally, (v) that of mass consumption. Rostow's stages are defined somewhat rigidly, universally and in economistic terms (that is, by the level of saving). Rostow's stages have been criticised as they do not take into account either the history of the new underdeveloped countries or their crucial relations with the new developed ones over several centuries past.

Some of the later formulations of sequence models have attempted to remove the rigidity and automatic character of the sequence by adding a word of caution that the evolutionary character of the sequence does not mean that each stage automatically generates its successor. Such evolutionary processes are not necessarily inexorable or inevitable. Taken with such caution and applied with greater specificity to the historical and structural context of development pertaining to each country situation, stages of growth hypotheses may be endowed with great normative significance, as this approach enables development to be guided in the prospective setting of a systematic structural transformation.

The idea of looking at the development process essentially as an evolutionary process of distinct stages and phases is also built into certain other approaches, for example, rural development. Since our main thesis in this work is that backward area development is essentially planning for transition growth, it is necessary to examine the various ideas that have been generated in this context.

Kulp[41] has distinguished four main stages in the agricultural development process in the rural areas. These stages are : (i) traditional subsistence agriculture, (ii) institution building development, (iii) institution based development, and (iv) capital intensive development.

Stage I is characterised by the subsistence orientation of farmers and the absence of accelerator institutions.

CHART I The Long-run Dynamics of the Rural Development System and its Environment (According to Kulp)

	Agricultural Development	Community Organisation	Individual	Planning & Financing	Project (Institutions & Innovations)
I.	Traditional subsistence Agriculture	Consultation and Small Project Management (Undertaking demonstration projects and self-help construction of infrastructure facilities at irregular intervals)	Peasant	—	Project (Institutions & Innovations)
II.	Institution-building Development	Infrastructure Management (i.e., Regular management of annual programmes	Businessman	Institutional & infrastructure development	Point 1
	II (a) Preparatory Pilot phase (Limited coverage phase)			Government assistance	1 1
	II (b) Coverage expansion phase				Line 1
III.	Institution-based development				1
	III (a) Completion of coverage by institutions & services—Beginning of diffusion of innovations	Business Management (Village leadership issues from project competence to business competence)	Entrepreneur	Loan financing becomes increasingly important	1 1 1 Network
	III (b) Momentum building-up phase	Service Management (The community business complex takes responsibility for providing a variety of accelerator services)		Bank credit	
IV.	Capital intensive Development				

In stage II, government and modern private institutions begin promoting innovations which results in a more intensive labour application. Some institutions expand their coverage to most of the villages. Most regions of developing countries are considered to be in this stage.

Stage III is considered the take-off stage when rural development acquires a self-sustaining momentum. In this stage, the rural institutional development is completed and farmers push labour intensification to the maximum. Rural income raises fairly steadily, discounting the vicissitudes of weather and prices.

In stage IV, investment and innovation turn to substituting capital for labour. Rural labour shortages appear; the price of farm labour goes up, and farmers apply capital to reduce the labour requirement per hectare.

Kulp also traces important psychological and sociological changes accompanying the above rural economic evolution in which the general trend is the assumption of more initiative for development by the rural community and the individual. He also notes that as the village economy passes through the four stages of agriculture development, planning and financing problems and procedures will also change substantially.

The basic entities of the rural development system are projects of three kinds namely, innovation, institution and infrastructure for which a three-stage progression is envisaged :

(i) Phase I : Point Phase—that is, trial in one place,

(ii) Phase II : Line Phase—that is, trial in several places under varying conditions, and

(iii) Phase III : Network Phase—that is, expansion at maximum speed as resources permit.

Thus, the rural development strategy, in essence, consists of directing the evolution of projects of innovation, institution and infrastructure along a path of gradualism—"Point-Line-Network", the pace and pattern of such development depending on the level of development of the rural community and its absorptive capacity.

The entire sequence of changes envisaged for rural development may be seen in the accompanying chart I showing the long-run dynamics of the rural development system and its environment.

STAGES OF GROWTH FOR SMALL COMMUNITIES—THE ANTHROPOLOGICAL APPROACH

The stage of growth hypothesis is also vigorously advocated in the planning of small tribal communities with low levels of technologies. Anthropologists who have studied tribal communities in different stages of development are of the view that, in the beginning, the tribal communities were in a state of ecological equilibrium with their environment. However, when this ecological equilibrium is disturbed for reasons either internal or external to the group, a stage of confrontation with other groups or institutions may result. For remedying this state of ecological disequilibrium in small communities, a four-stage model has been advocated.[42] The stages envisaged are as follows :

Stage I. State of confrontation

Stage II. (a) Countering external forces—institutional ;
(b) Institutionalising measures for countering individual or group forces ;
(c) Effective attention to problems of health ; and
(d) Recognition of the native skills, restoration of self-confidence and citizen education.

Pause

Stage III. (a) First-aid to the traditional economy
(b) Breaking the communication barriers and education
(c) Gradual upgradation of native skills and introduction of new elements in a longer time-perspective.

Confrontation ceases

Stage IV. The group is free to share the experience of larger communities.

HYPOTHESIS CONCERNING CONVERGENCE AND DIVERGENCE OF REGIONAL INEQUALITIES

Among the empirical studies relating to convergence and divergence in regional income differences in relation to economic growth, Williamson's[43] study is an important contribution. Williamson's analysis covers twenty-four countries including the Philippines and India which have been selected for case studies in the present work. He concluded that both cross section and time series analysis indicated that there was a systematic relation between national development levels and regional inequality. According to him, regional income differences will initially diverge under the conditions of economic growth, but will decrease after the country achieves a certain level of development. "Rising regional income disparities and increasing North-South dualism is typical of the more mature stages of economic national growth and development".[44] Williamson also emphasises that he has made no attempt "to measure regional concentration of population and urbanization", but has only measured "regional variation in per capita income levels, weighted by the distribution of population".[45]

The conclusions of Williamson have been questioned by some while corroborated by others and they still constitute a lively subject of debate. Gilbert and Goodman[46] have pointed to the lack of representativeness of the countries selected and especially to the arbitrary delimitation of regions within these countries. In contrast to Williamson, these authors conclude that convergence does not occur and they even question the same possibility of convergence in underdeveloped countries. In their opinion, the degrees of income differences between regions coincides with increased socio-economic inequality within regions. Friedman pointed out that Williamson's conclusions are only partly borne out by detailed historical studies. According to him, "the best that can be said from the available information is that regional income convergence is an extremely drawn-out process. It occurs, when it does, at a relatively advanced stage in the industrialisation process, and it rarely changes the rank-ordering of regions by income".[47]

Slater[48] in his analysis of Brazil, Pakistan and the Ivory Coast comes to similar conclusions. The investigations by Mera on the Japanese and Korean cases provide evidence of convergence of per capita income with a generally high rate of national economic growth. From this, he concluded that "it appears quite probable that the maintenance of high growth of the economy is an effective way of reducing the rate of population concentration as well as reducing income disparities among regions, once the economy reaches a certain stage.[49] Explaining the convergence in prefectural per capita incomes, Mera emphasises two reasons : (i) the convergence in wage rate differentials among different parts of the economy which has in turn been caused by the lightening of the labour market and (ii) the improved mobility of people and better information made possible by improvement of transportation and communication services.[50] However, he points out that a high rate of economic growth leading to equalisation of income and wage levels within the country is a process which may take a long time. In the case of Japan, the process described by Mera took about 30 years, but according to him it should require less time in developing countries, if these countries can maintain a high rate of national economic growth, It should be noted in this context that Japan's case must be considered atypical, although it belongs to non-western economic development experience.

Hansen[51] in his critical review of Williamson's thesis points out that the hypothesis of convergence of disparities with economic growth "ignores structural issues and implicitly adopts the view that macro-economic policy is all that matters; regional disparities—and presumably other structural problems—will then take care of themselves." Pointing to the fact that this assumption is simplistic and that the situation in developing countries is more complex than in developed countries, he observes that "issues involving human resource development and basic values may need to receive primary attention before regional problems can be successfully attacked either directly or indirectly, through sustained national economic growth. By yet another way, it may be necessary to deal with fundamental structural problems before national growth can proceed to a point where it positively affects remaining structural problems."

A recent study by Chenery *et al.*,[52] provides some insights into the relations between growth and distribution. According to the authors, there is evidence that in the early stages of development, the distribution of income tends to become more concentrated. "Increases in output come disproportionately from relatively small modern sectors of primary production and industry, which absorb a high proportion of total investment and have relatively high rates of productivity growth. This pattern of concentrated growth is perpetuated by limited access to land, credit, education and modern sector employment and is often reinforced, unintentionally or otherwise by the government's fiscal and trade policies as well as the distribution of public expenditures. As growth continues, its benefits are more widely spread, but there are a number of obstacles that limit the share received by the poor".[53] The contention of these authors is that although growth tends initially to be concentrated in a few sectors of the economy, with little effect on major poverty groups, it is possible to devise suitable policies (as some countries have shown) for offsetting this tendency so that the benefits of growth can be shared more equally. Adelman and Morris[54] have analysed economic and political data from 74 underdeveloped countries for the period 1957-68, with a view to understanding the extent to which benefits of economic growth in underdeveloped countries reached those most in need. They used statistical-mathematical analyses of social, cultural, economic and political variables (including among such variables a broad spectrum of qualitative phenomena) to trace those variables showing a close relation with differences in income inequality between and within the countries selected for their studies. They conclude that "inequality of income tends to be greatest where the exploitation of an abundance of natural resources coincides with a concentration of assets in the hands of expatriates; it tends to be least where development strategies stress investment in human resources, greater diversity of manufacturing exports and expansion of public sector output and investment".[55] Their studies support the Marxian view that economic structure, not level of income or rate of economic growth is the basic determinant of patterns of income distribution. Their studies also suggest that there is no automatic, or even likely, trickling down of the benefits of economic growth to the poorest segments of the population in low-income countries. "On the contrary, the absolute position of the poor tends to deteriorate as a consequence of economic growth".[56] The importance of social, institutional and political dimension of development "for the achievement of greater political and economic equality" is stressed in this contribution, that is, aspects which do not enter into the discussion on economic growth models.

THEORIES OF SPATIAL STRUCTURE BASED ON MARXIAN ANALYSIS

An interesting departure from the abstract empiricism and perspectives of the modern Anglo-Saxon geography of underdevelopment is provided by another group of contributors who have provided interpretations based on Marxist Doctrine and Analysis. Among these 'neo-Marxist' group of writers, the work of the Latin American School deserves great attention. The concepts of "internal colonialism", "peripheral capitalism", "marginalisation" and "dependency economics" belong to this school. The Latin American experience has shown that with the penetration of the capital system, all kinds of polarisation—social, spatial and economic—emerged and increased. As a result, the Latin American economies became not merely "peripheral" but also "dependent". This dependency had both internal and external structure. According to Frank, this process operates both internationally and internally within countries. "Economic development and underdevelopment are relational and qualitative, in that each is structurally different from, yet caused by its relation with the other. Yet development and underdevelopment are the same in that they are the product of a single, but dialectically contradictory, economic structure and and process of capitalism... one and the same historical process of the expansion of capitalism continues to generate—both economic development and structural underdevelopment"[57] These are very relevant to our context, as they are rooted in empirical reality. As Brookfield[58] remarks: "The particular interest of the Latin American School from this point of view is its base in empirical reality, for the new thinking has not arisen through the slavish adoption of external models, but rather through questioning the relevance of these models, to the historical and actual experience of the region."

In the distinctive pattern of development thinking evolved among the Latin American scholars, three main directions may be seen: First, an analysis of trade relations among countries on the basis of centre-periphery imbalance, most specifically in regard to the terms of trade ; second, a strong emphasis on industrialisation with the objective of rectifying this imbalance; thirdly, and somewhat later, a concentration on the social and structural problems of underdevelopment. Among the Latin American scholars who advanced this kind of development thinking, there are politicians, academics and practising economists. The basic ideas of this thinking had been carried forward by a number of others in different parts of the world who have applied this doctrine with some variations to different country situations. Thus we have today a whole school of thought based on the Marxist theory of inequality. The contributions[59] of Mariategui, Furtado, Sunkel, Stavenhagen, Beckford, David Slater, Casanova, Girvan, Kenneth, Griffin, Samir Amin, Richard Peet, Harvey, Anderson, Folke, Shivaji and Chattopadhyaya and Raza may be particularly mentioned in this context.

According to Samir Amin, history has seen three sets of peripheral formations: (a) The Latin American, (b) Arab and Asian, and (c) African. Within each set, there are numerous variations. Despite all these differences, all the peripheral formations tend to converge toward a pattern that is essentially the same. Amin claims that all peripheral formations have *four* main characteristics in common (i) the predominance of agrarian capitalism in the national sector, (ii) the creation of a local, mainly merchant, bourgeoisie in the wake of dominant foreign capital, (iii) a tendency toward a peculiar bureaucratic development, specific to the contemporary periphery; and (iv) the incomplete, specific character of the phenomenon of proletariarisation.

Based on historical analysis, Amin has also outlined a theory of the transition to peripheral capitalist economy which contains nine hypotheses. A summary of these nine hypotheses may be stated as follows :

(1) The onslaught from without, by means of trade, carried out by the capitalist mode of production upon the precapitalist formations, causes certain crucial retrogressions to take place, such as the ruin of the crafts without their being replaced by local industrial production.

(2) Unequal international specialisation is manifested in three kinds of distortion in the direction taken by the development of the periphery, namely distortion toward export activities, reduction in wage rate and the consequent unequal exchange, and the subsequent pattern of industrialisation through import substitution.

(3) This initial distortion brings another in its train, the hypertrophy of the tertiary sector in the periphery, which neither the evolution of the structure of demand nor that of production can explain. This hypertrophy of unproductive activities, expressed especially in the excessive growth of administrative expenditure, is manifested in the Third World of today by the quasi-permanent crisis of government finance.

(4) Unequal international specialisation also underlies the distortion in the periphery toward light branches of activity, together with the employment of modern production techniques in these branches.

(5) The theory of the multiplier effect of investment cannot be extended in a mechanical way to the periphery.

(6) Analysis of the strategies of foreign monopolies in the underdeveloped countries shows that the periphery is without economic means of action in relation to the monopolies.

(7) Underdevelopment is manifested not in the level of production per head, but in certain characteristic structural features that oblige us not to confuse the underdeveloped countries with the now-advanced countries as they were at an early stage of development.

(8) The accentuation of the features of underdevelopment necessarily results in the impossibility of going over to autocentric and autodynamic growth.

(9) And finally, while at the centre the capitalist mode of production tends to become exclusive, the same is not true of the periphery. Consequently, the peripheral formations are fundamentally different from those of the centre.

Generally speaking, the Neo-Marxist group of writers seeks to explain the nature of underdevelopment and spatial organisation under colonialism that is, in colonial states charac erised by the development of unified political control over the process of production. In these countries, exploitative form of relationships established which led to extreme spatial inequalities in development. Slater refers to two fundamental spatial tendencies in this process which he characterises by the terms territorial centralisation and spatial penetration. By territorial centralisation, he means the specific and accentuated form of spatial concentration which is the product of the accelerated aggravation of the means of industrial production and the commodity of labour power in one major urban agglomeration. By spatial penetration, he means the process by which both international and local capital, but specially the former, is being extended into interior zones in order to acquire raw materials and agricultural resources and made to fuel the world-wide process of expanded capital accumulation.[75]

It is noted that an important aspect of the spatial structure in these countries is the marked unevenness in the levels of development and the productive forces and the conditions under which surplus is generated and realised. Usually the objects of exploitation are the primary agricultural and mineral resources that are in demand in the metropolitan economies, so that regions possessing particular mineral resources such as gold, copper or lead experienced capital penetration at a relatively early stage. Another characteristic of development in the colonial countries is that the technology of production in the zones of capitalist penetration stands in sharp contrast and divorced from indigenous technology. Since production is monopolised by externally based companies, there is no gradual internal integration of more advanced methods of production within the capitalist zones of the colony. Further, in the colonial situation, the wage labour was low and did not induce any internal demand for mass consumption goods. As the areas that supplied labour for capitalist production were denuded of much of their available manpower, their own economic self-sufficiency in these areas was often undermined. This process further increased the spatial dichotomy.

In the areas characterised by capitalist relations of production, the surplus that is generated is controlled and appropriated by a metropolitan economy located within the geographic confines of the colonial countries. In other parts of the colonial territories, the relations of production will be different. In some areas, semi-feudal relations of production may still prevail, and yet again in other areas, varying forms of traditional communalism will occur, where the contours of socio-economic differentiations have not emerged in any sharply defined way.

In the colonial areas, it is the manner of the appropriation and utilisation of the surplus by particular social classes that becomes the key component in understanding not only underdevelopment as a whole but also its associated spatial structure. Depending on the nature of the exploitation by the colonial power, the form that the surplus takes will also vary. According to Shivji, three such forms can be distinguished namely, (i) surplus value, (ii) merchants' profits, and (iii) surplus labour. For example in some zones, where one has plantation-based agriculture, mining, quarrying and small-scale manufacturing, the surplus largely takes the form of surplus value which is extracted by the owners of the means of production. In another area, where peasant-based agriculture and small commodity production are the predominant features, the surplus is extracted through unequal exchange, taking the form of merchant profits. Again, in non-monetized zones, surplus may be extracted either through the provision of free (surplus labour) for example bonded labour or through the supply of food to the zones of capitalist production.

In the colonial economy, a substantial part of the surplus that was extracted was drained away to the economy through repatriation of open and hidden profits of the foreign owners of the means of production and through unequal exchange of the commodities with world market. A disproportionate percentage of the surplus that remains in the colony is utilised in the non-productive services sector such as, commercial and financial infrastructure and also in public administration. Some of the retainee surplus may be invested in the export-oriented agriculture sector, but essentially the accumulation takes the form of commercial capital and is not converted into industrial capital.

The vast bulk of the retained surplus is utilised in :

(i) the zone of export production in order to ensure the expanded production of future surplus in these areas ;

(ii) the various urban places that act as administrative and commercial centres within the colony ; and

(iii) in the construction of transport and communication link between the different points and zones of the export economy.

Thus in a colonial economy, the spatial structure which evolves is very much externally oriented with links between the internal regions of the colony being very limited indeed. The unfortunate trend is that even after the end of colonialism, that is, even in the period after independence this pattern of underdevelopment persists. Slater adduces five reasons for this persistence. First, much emphasis continues to be given to export production, especially in agriculture, and in order to increase the "efficiency" of this production, considerable investment is constantly required. Secondly, in those rural areas, previously characterised by the rise of an indigenous capitalist class, investment also tends to be made, because of the greater possibility of an adequate return on the capital invested. Thirdly, the industries that are established, being in the main either export processing or import substituting in nature, tend to be located in either the areas of export production or urban centres, and especially in the largest urban agglomeration, where one has the greatest concentration of already existing industries, a skilled and semi-skilled labour force, a ready market for the goods being produced, superior infrastructure and services, access to a major port and so on. Fourthly, aid provided by international capitalist agencies such as the World Bank, gravitated towards the major urban centres, for example into modern infrastructure projects or into areas of export agriculture such as tea and coffee production, because of the greater likelihood of profitable returns in the already "developed" areas. Fifthly, and very much related to the last two points, foreign capital also prefers to locate new factories in already semi-industrialized areas, and conversely projects in backward zones are quite naturally regarded as much more hazardous economically.[60]

Therefore, in the view of this group of writers, the very structure of underdevelopment has to be radically altered to bring about a spatially egalitarian structure of development. How to bring about such transition in different types of economies is not specifically answered by most writers. Most of them seem to be concerned mainly with exchange relations (in international trade with advanced capitalist countries) rather than the relations of production.[61] Thus they emphasize that both the nature of the relationships between the under-developed economy and the world capitalist system and the internal alignment of class forces within the country must be changed. Such changes have been carried through in countries like China, Cuba, North Korea and North Vietnam, and in all these cases, the organisation and structure of space has been correspondingly altered in various fundamental ways.

Sociological contributions

Our discussion of development thought centring around the theme of inequalities will be incomplete without a consideration of the sociological stream of thought. This stream abounds in theories on poverty, relative deprivation, opportunity structures, social well-being and income distribution. The discussions on these lines provide a rather different perspective with their implications for economic and social policies. They have led to the formulation of new development strategies for the developing countries among which anti-poverty strategies, basic needs strategy and employment-oriented strategies may be specially mentioned.

Most of the discussions in the sociological stream centre around the theme of poverty. Before we proceed further on the subject, it is necessary to distinguish at least in theory the distinction between poverty and inequality. It is true that to some extent, poverty and inequality are inextricably linked. The concern with inequality is frequently presented in the form of statements about poverty. Also, efforts to reduce poverty often require reductions in inequalities ; this is probably the case of educational differentiations in societies with rising educational requirements. But not every statement about inequality is necessarily a statement about poverty. However, it may be noted that in developing countries poverty and inequality are closely linked.

Poverty and backward areas

There is a close affinity between poverty which is an inter-personal concept and backward areas which is an inter-regional concept. Both are relative concepts. Apart from this resemblance, the depressed regions in the developing countries are also coterminous with areas of hard-core poverty intertwined with the profile of backwardness. It is difficult to separate the two, as the one becomes the cause of the other. In all these cases, the underdevelopedness, by and large, refers to the state of being of the peoples, their culture patterns and their levels of living. Thus the terms like traditional economy, subsistence economy and tribal economy used with reference to some backward areas in the developing countries refer to the association of culture patterns and levels of living with area backwardness, which is usually defined with reference to the level of utilisation of the natural resources and infrastructure endowment of an area. Thus the development of backward areas in the developing countries is not merely a question of transferring technology or building up infrastructures. It has many other critical elements besides, such as the nature, level, quality and internal stimulus of the culture of the inhabitants of the areas, that is, characteristics usually embedded in the social, political and religious structure and philosophy of the people, their attitude towards change in social and material culture and their innate ability to cope with the problems of the area.

According to Galbraith[62] poverty tends to beget certain attitudes and behaviour that make it self-perpetuating. Thus conservatism, the limited ability to undertake a measure of risk in anything, illiteracy and the limited horizons it implies, the inertia resulting from poor health and malnutrition, the biological consequences of high birth-rate and death-rate, the poor savings and ability to invest, unemployment, the lack of social institutions and skilled manpower to make effective use of capital are some of the characteristics that go with poverty and therefore with areas with concentration of the poor. That is why the strategies being evolved in many developing countries are not merely focusing on the development of backward *areas* only but are also seeking to relate themselves to the uneven advance of *individuals* or groups in the area that is, poverty.

Usually the poverty and cultural dimensions of the backward area problem are not captured adequately in the specific economic statistical indices used for planning in the developed countries. That is why planning for the underdeveloped areas in the developing countries calls for indigenous appropriate methodologies for tackling their problems.

As may be seen in this context, the perception of poverty and its relationship to the backwardness of the area is an essential step in the planning for the backward areas in developing countries. Therefore, the discussion on poverty is a very relevant issue in the geography of under-development.

In the sociological contributions on poverty four major themes repeatedly appear namely, (1) poverty is a social phenomena, (2) it must be related to a study of the total society, (3) it can be seen as either relative or absolute, and (4) programmes to alleviate poverty must take into account its heterogeneity. We may develop these themes further and discuss some concepts which are very much relevant to our present discussion.

The 19th century thinkers, particularly the Social Darwinists[63] saw poverty as the inevitable outcome of the struggle for survival. The principles of natural selection and survival of the fittest were fundamental laws of social development which invoked a seemingly indisputable biological premise to social inequalities. The early twentieth century sociologists, however, recognised that poverty was primarily a social and not a biological consequence. Thus poverty came to be defined as a culture. According to Cooley,[64] "poverty is unfitness, in a social and not a biological sense" and also that poverty is an unintended consequence of economic development. Much later, the social inequalities which accompanied the industrialisation process in developed countries led many authors to see poverty as an outgrowth of the structural differentiation that occurs in an industrial society. This led to the further idea that poverty is relative deprivation.[65] Relative deprivation implies a social context and a set of normative standards which served as a basis for comparative evaluation. According to Townsend "poverty must be regarded as a general form of relative deprivation which is the effect of the maldistribution of resources" and "that sections of the population whose resources are so depressed from the mean as to be deprived of enjoying the benefits and participating in the activities which are customary in that society can be said to be in poverty".[66]

The above inequality theories like culture of poverty and relative deprivation have been used as the basis for anti-poverty policies in most western countries. However, their relevance to the developing countries has been very much questioned. In developing countries, particularly in democratic societies, amelioration of poverty essentially means ensuring equality of access to the rewards of development, as well as equality of distribution of awards for various groups. In this context, the concept of opportunity structures is considered to be important. It has been contended that differences in influence, status and material wealth are the result of social structures of a more permanent character, entailing the unequal access of individuals and groups to the material and non-material resources of a society. Therefore, it is the inequality in the social structure which leads to unequal development. Thus, it is the access to opportunities that has been identified more with the inequalities problem in developing countries. This may be explained from the fact that in an agrarian society, for example, access to land will be a more important determinant of income inequality than in an industrial society in which perhaps access to education may be a crucial factor. Thus, equalitarianism in the sense of equality of opportunity is increasingly being defined as a value and a goal in the plan documents of many developing countries. In the specific anti-poverty strategies which are emerging in the developing countries, two major approaches are to be considered, namely, (1) social conditions approach and (2) the social mobility approach.

The social conditions approach places emphasis on programmes of improving *social conditions* by attempting to change the distribution of benefits. In this context, five variants of the efforts to improve social conditions may be mentioned. (i) Improving the income levels of a substantial category of the poor directly such as, through a minimum wage.

Employment programmes through public works resorted to in many developing countries is an example of this type of programme, (ii) Insuring against economic stress and old age for example, old age pensions, unemployment insurance etc, (iii) Pinpointing aid to particular groups who are sharply falling behind the general advances of society through supplementary and assistance schemes to such population, (iv) Provision of basic needs, (v) Enhancing social conditions to promote social mobility.

Closely allied to the social conditions approach to poverty is the concept of social well-being which has been elaborated by Smith[67] in the United States.

Smith has established seven general criteria of social well-being relevant to contemporary conditions in the United States. For developing countries, similar criteria can be established. In fact, the basic needs approach which is increasingly coming into prominence in the policy/strategy formulation of many developing countries is a close parallel to this approach.*

The social mobility approach, in contradistinction to the social conditions approach, seeks to change the distribution of individuals by improving their chances to move into occupational niches which provide higher levels of income and amenities. The social mobility approach is highly valued because it emphasises the normal operation of market forces and the desirability of participation in the labour market. Economic development means to a large extent social mobility,

*The importance of adequate levels of nutrition, clothing, shelter, health and other physiological requirements for survival is now universally recognised. To these basic needs, some cultural needs of education, security and leisure together with certain qualitative aspects of the physical, social political and economic environments have to be added. Thus, most of of the components of social well-being are ultimately based on categories of need. Such need may be defined in a normative manner by establishing survival standards. This is, referred to as normative need, absolute need or basic need. This is a kind of approach relevant for the present to most developing countries, where poverty is conceived in terms of subsistence criteria. As opposed to this, in developed industrialised societies, the need is equated with one of latent demands that is interpreted as a feeling of relative deprivation.

for it requires and results in changes in the occupational distribution. From the viewpoint of policy, the improvement of social mobility chances for the poor would require an expansion and democratisation of education. In developing countries, it may be asserted that imparting of skills and non-formal education may be regarded as key issues—in the social mobility approach to poverty alleviation.

The social mobility approach is oriented to opening "doors" into the larger society for the poor; the social conditions approach, to providing "floors" so that no one can fall below a certain minimum level. The two approaches obviously overlap. But they are aimed at different visions and values of society and different concerns about the poor.

To what extent do these approaches necessarily conflict, to what extent do they converge, and which one has precedence in what kinds of situation are questions on which we do not have adequate answers today for the developing countries.

Some antitheses to growth-oriented strategies

It is clear from several accounts that the developing countries are characterised by a highly uneven character of development. It is also clear that the development interventions attempted during the 50s and the 60s (based on growth theories), have not led to any significant improvements in this regard, but may have, if anything, only increased inequalities and accentuated the existing problems. In our review, we have noted that growth-oriented strategies have tended to neglect many issues that are of great relevance to the developing countries.* Among them are: redistribution of incomes, employment, the structural constraints to labour-absorbing growth such as the lack of entrepreneurial ability and skilled technical and managerial manpower; and structural changes in the social,

*Kenya Employment Report, for example, has postulated an overall growth rate of 7% per annum as being necessary for the implementation of a development strategy based on employment creation and income redistribution—See ILO : Employment, Incomes and Equality: A Strategy for increasing productive employment in Kenya, Geneva, 1972.

political, institutional and value systems. On account of these serious omissions, the conventional wisdom of growth-oriented strategies have been seriously questioned from the late' 60s onwards.

The disillusionment with growth-oriented strategies have led some thinkers to advance arguments that are antithetical to growth. Thus antigrowth, antiaffluence and antidevelopment hypotheses have been advanced.

The following statement may be considered typical of this school of thinking :

> Development is also not a desirable ideal. For development can work only when inequity as a value, is enshrined in the society. If all people have cars in Varanasi,* for example, nobody will be able to use a car in the city. Development is then functional only when a few people have access to it. In a modern society, where one pleads for equality, where the resources are limited and their indiscriminate, unsparing exploitation leads only to untimely exhaustion-liberation lies not only in 'limited' growth, but in the limitation of wants, in lowering down the standards of living of the few rich and powerful as also in the achievement of 'no poverty', no 'development' society. The alternative plan for social change, the antithesis of development, will lead us to the latter in the shortest possible time. It will help liberation of all.[68]

It has been argued that development has been disastrous for civilisation, has divided the world into haves and havenots and therefore a contentment-orientation is necessary for the survival of man. According to these thinkers, the New Society will produce goods and services that will be determined by the needs and resources of the people. The productive process of a country will then provide only such goods as are required for the fulfilment of the basic needs of all its citizens. The New Society will devote itself to the non-exploitative means of production. It will be equity oriented. A new technology

*Varanasi is a city in India which is noted for its narrow streets and lanes.

which will be appropriate to the productive process will become an operational measure for the establishment of the New Society. The New Society will create such an employment system which will be essentially productive, but not organised by a few for the rest, but by those who need employment. The goal of the 'No Poverty' society is not to enrich a few, or even the many, but to help all. The aim is not to achieve the greatest good of the greatest number, but the uplift of everyone, the good of all. Such a society is described is *Ujamaa* by Nyerere and as *Sarvodaya** by Vinobha in India and Aryaratne in Sri Lanka (*Sarvodaya* means "awakening of all"). These are some of the movements that have grown in the developing countries with the aim of fostering social reconstruction through self-reliant methods.**

The heterogereity in the cultural and economic settings of the different countries of the developing world seems to suggest that varied solutions will perhaps have to be generated to meet the common problems of development. There is no one best

---

* The Sarvodaya Shramdan movement of Sri Lanka has four clear objectives of awakening viz. the awakening of personality of every single individual—man, woman or child ; awakening of village community: awakening of national community ; and the awakening of the world community. The philosophy has identified two sets of factors as constraints to development, one within the individual, the village or the nation and the other from without. If awakening is to take place, both sets of constraints have to be removed. For example, in case of a village, various divisible forces and exploitative processes within the village itself, springing from mutual distrust have to be removed, while at the same time those forces from outside the village that keep the village in bondage and dependence such as absentee, land-owners, money-lenders, middlemen and traders, also have to be eliminated. Similarly, at all levels these forms of constraints have to be counteracted and removed. For details see, *Sarvodaya Shramdan Movement for Social Development in Sri Lanka : A study of experience in generating people's participation*' by A.T. Aryaratne—Working paper for United Nations Children's Fund—Special meeting on the situation of children in Asia with emphasis on basic services, United Nations, E/ICEF/ASIA/8, dated 19th April, 1977.

**The indicators of growth in this economy is not to be determined by the GNP; but by the rate at which the weakest member of the society grows, develops and is finally rescued from poverty.

path to development. We may therefore divert our attention to some indigenous development models.

A development model with man, not growth as the prime objective is the Gandhian model, which outlines a complete system of organic growth of human society. The basic philosophy of the Gandhian model is contained in the following statement made by Gandhi :

> According to me, the Economic Constitution of India and for that matter that of the whole world should be such that no one under it should suffer for want of food or clothing. In other words, every one should be able to get sufficient work to enable him to make both ends meet. And this ideal can be realised only if the means of production of the elementary necessities remain in the control of masses.[69]

Gandhiji was not against industrialization per se. He however, wanted the size of industries to be within manageable limits :

> As a moderately intelligent man, I know that men cannot live without industry. Therefore, I cannot be opposed to industrialisation. But I have a great concern in introducing machine industries. The machine produces much too fast and brings with it a sort of economic system that I cannot grasp. I do not know to accept something when I see that its evil effects outweigh whatever good it brings with it. I want the dumb millions in our land to be healthy and happy. I want them to grow spiritually. As yet, for this purpose we do not need the machine. There are too many idle hands. But as we grow in understanding, if we feel the need for machine, we certainly have them, once we have shaped our life on 'Ahimsa', we shall know how to control the machine.[70]

Gandhi's emphasis lay not on technology but on man—his organization, his value system and his ethics. To him, the machine was only a tool ; technology was only a means to an end, which is the development of man. Technology was not something outside the society. Society produced technology and technology produced society in an endless mess of action and interaction. According to Gandhi, one cannot override the other.

The Gandhian model is basically rooted in the removal of inequalities, both interpersonal or group and spatial. His concern for spatial inequalities may be inferred from his dislike for the growth of big cities. On the question of human settlements, Gandhi was in favour of villages and did not like the giganticism of cities. He wanted each Indian village to be self-sufficient in its basic needs—food, shelter and clothing.

There is a lot that is essentially human in the gandhian model. It answers well the oft-repeated question : "Growth for What ?" The gandhian model is highly integrated and organic. It is progressive, humane and synthetic. According to Gandhi, real development is one which is organic in the sense that growth and development of each element of the social system has not only to be rapid, but more important than that, it should be also proportionate. Excessive growth of a single element is cancerous. And that is what modern society suffers from.

The gandhian model or other social construction models like Sarvodaya contain many indigenous elements, that is, elements which are antithetic to the Western inspired growth models. They are, however, country-specific, both in origin and development, and are therefore circumscribed by the conditions prevailing in their own countries. They have some constraints for universal application.

NOTES AND REFERENCES

1. According to Schumpeter, by development is understood "only such changes in economic life as are not forced upon it from without, but arise by its own initiative, from within". The essence of development is the emergence of qualitatively new phenomena within the economy which may include : (1) The introduction of a new good—or a new quality of a good; (2) The introduction of a new method of production ...which need by no means be founded upon a discovery scientifically new, and can also exist in a new way of handling a commodity commercially; (3) The opening of a new market...whether or not this market has existed before; (4) The conquest of a new source of supply

of raw materials or half-manufactured goods. . . (5) The carrying out of the new organisation of any industry, like the creation of a monopoly position. . . or the breaking up of a monopoly position.

2. R. Nurkse, The Theory of development and the idea of balanced growth in A.B. Mountjoy, *Developing the underdeveloped countries*, Macmillan, 1971. Also Rosenstein – P.N. Rodan, Problems of Industrialisation of Eastern and South-eastern Europe, *Economic Journal*, 1943, Vol. 53, pp. 202-211; R. Nurkse, Some international aspects of the problem of economic development, *American Economic Review* 1952, Vol. 42, pp. 571-583.
3. M. Fleming, External economy and the doctrine of balanced growth, *Economic Journal*, Vol. 65, 1955, pp. 241-256.
4. The polarisation effect refers to the gradual deterioration of the backward region due to competition and outmigration of capital and labour (including brain drain).
5. These trickling down forces work particularly through inter-regional trade and transfer of capital to the backward regions. Their effect depends largely on the existence of complementaries between the industries in the growth centre and the hinterland. In addition, migration from the hinterland to the growth centre may absorb some disguised unemployment and raise the marginal productivity of labour and the income per capita of the hinterland.
6. G. Myrdal, *Economic Theory and Underdeveloped Regions*, London, 1957.
7. F. Perroux, Economic space : Theory and application, *Quarterly Journal of Economics*, February 1950, also reprinted in J. Friedman and W. Alonso, *Regional Development and Planning : A Reader*, Cambridge, Mass, 1964.
8. Ibid.
9. Ibid.
10. J.R. Boudeville, *Problems of Regional Economic Planning*, p. 11.
11. op. cit., p. 112.
12. T. Hermansen, "Development Poles and Development Centres in National and Regional Development : Elements of a Theoretical Framework" (mimeographed).
13. J.R. Boudeville, op. cit., p. 112.
14. A.J. Youngaon, *Possibilities of Economic Progress*.
15. P.N. Rasmussen, *Studies of Intersectoral Relations*, 1956.
16. This is said to have "drained the growth pole concept of its original temporal and dynamic meaning and recharged it with a static and/or comparative static content", J.R. Lasuen, On growth poles, *Urban studies*, Vol. 6, No. 2.
17. N.M. Hansen, (Ed.), *Growth Centres in Regional Economic Development*, New York, 1972.
18. J.R. Lasuen, On growth poles in Hansen (Ed.), *Growth Centers in Regional Economic Development*, New York, 1972.
19. R.F. Darwent, *Growth Poles and Growth Centres in Regional Develop-*

ment : A Review, Environment and Planning, Vol. I, 1969.

20. J.R. Lasuen, A generalisation of the growth pole notion, paper for the International Geographical Association meeting, Victoria, 1971.
21. T. Hagerstrand *Innovation Diffusion as a Spatial Process*, Chicago, University of Chicago Press, 1953.
22. L.A. Brown, *Diffusion Process and Location : A Conceptual Framework and Bibliography*, Regional Science Review and Bibliography series No. 4, Philadelphia, 1968.
23. R.L. Morril, and F.R. Pitts, "Marriage, Migration and Mean Information Field", *Annals of the Association of American Geographers*, Vol. 57, pp. 401-22.
24. J.R. Lasuen, "Urbanisation and development : A temporal interaction between geographical and sectoral clusters", paper to the Conference on growth pole hypotheses and policies, Madrid, September, 1970.
25. R.P. Misra, K.V. Sundaram and V.L.S. Prakasa Rao, *Regional Development Planning in India : A new Strategy*, Vikas, New Delhi, 1974.
26. J.L. Brian Berry, Social change as a spatial process, *International Social Development Review* No. 4, United Nations ,New York, 1972.
27. J.R. Lasuen, Ibid.
28. P.R. Gould, Tanzania, 1920-63 : The spatial impress of the modernisation process, *World Politics*, Volume 22, 1970, pp. 149-70.
29. J.B. Riddell, *The Spatial Dynamics of Modernisation in Sierra Leone, Structure Diffusion and Response*, Evanston, III. 1970.
30. Edward Soja, *The Geography of Modernisation in Kenya : A Spatial Analysis of Social, Economic and Political Change*, Syracuse, New York, 1968.
31. E.W. Soja, and R.J. Tobin : The Geography of Modernisation, paths, patterns, processes of spatial change in developing countries, in R. Brunner and G. Brewer (Eds.), *Ordered Complexity : Empirical theories of Political Development*, New York, 1972.
32. Riddell's argument is significant : "cultural and ethnic patterns within the population are very real and do influence alterations, but the overall pattern of change in the country is a function of its geography—The spatial organization of the countryside by network and hierarchy. The spatial dimension affords the key focus toward understanding the phenomenon of modernization" Ibid. p. 131.
33. J.L. Brian Berry, Social Change as a Spatial Process, *International Social Development Review*, No. 4, United Nations, New York, 1972.
34. P.S. Towfighi, Strategies for regional growth in developing countries, *Human Settlement*, July, 1972, Centre for Housing, Building and Planning, United Nations, New York.
35. D.C. North, Location theory and regional economic growth, *Journal of Political Economy*, Vol. 63, June, 1959.
36. J.G.N. Hilhorst in his analysis of the Economic Base Theory concludes that two conditions should be fulfilled for regional growth to be successful viz. (i) There should be the economic possibilities for diversification, and (ii) The socio-political structure should allow for

it. See J.G.N. Hilhorst, Regional development theory, an attempt to synthesise in *Multidisciplinary aspects of regional development*: OECD, Paris, 1969.

37. Perloff, *et al.*, *Regions Resources and Economic Growth*, Baltimore, 1960.
38. J.G.N. Hilhorst, Regional Development Theory, an attempt to synthesise in *Multi-disciplinary aspects of regional development*, OECD, Paris, 1969. Hilhorst remarks, "The Theory fails in its strive for generality. . . . .The theory takes on too much of a mechanistic character as it requires a region to go from one stage to another, and does not allow for jumps".
39. M.H. Watkins, "A staple theory of growth, *Canadian Journal of Economics and Political Science*, Volume 29, 1963. Also R.E. Caves, Export-led growth and the new economic history in J.N. Bhagwati *et al.* (ed) *Trade, balance of payments and growth* : *Essays in International Economics in honour of Charles P. Kindleberger*, North Holland, Amsterdam, 1971.
40. W.W. Rostow, *The Stages of Economic Growth*, Second Edition, New York, 1971.
41. E.M. Kulp, *Rural Development Planning : System Analysis and Working Method*, Praeger, New York, 1970.
42. B.D. Sharma, Development of Small Tribal Community (Mimeo) Applicability of the "Stages of Growth" hypothesis to small communities is further discussed in the chapter on India.
43. J.C. Williamson, Regional inequality and the process of national development : A description of the patterns, *Economic development and cultural change*, vol. 23, 4, 1965, pp. 3-45.
44. Op. cit., p. 44.
45. Op. cit. p. 16. Coates *et al.* underline this point to explain the presence of regional problems even in U.S.A. and U.K., the world's mightiest economic nation and the world's oldest industrial state respectively. They drive home the point that convergence of regional per capita income may not, in fact, be accompanied by regional convergence of opportunity, economic power and quality of life. See Coates *et al*, *Geography and Inequality*, Oxford, 1977.
46. A. Gilbert, and D. Goodman, Regional income disparities and economic development, a critique in A. Gilbert (Ed.), *Development Planning and Spatial Structure*, London, 1976. pp. 113-141.
47. J. Friedmann, Urbanisation, Planning and National Development, Aage, 1973.
48. D. Slater, Underdevelopment : Spatial inequality, approaches to the problems of regional planning in the Third World, Progress in Planning, Volume 4, part 2, pp., 97-167.
49. K. Mera, The changing pattern of population distribution in *Growth Pole strategy and regional development planning in Asia.*, United Nations Centre for Regional Development, Nagoya Japan, 1976, pp. 242-277. Also, K. Mera, Population concentration and regional

income disparities : A comparative analysis of Japan and Korea in N. Wansen (Ed.), *Human Settlement System : International Perspectives on structure, change and public policy*, Ballinger, 1977, pp. 155-176.

50. A third plausible reason for the convergence of per capita incomes may be the reserving scale economies or even diseconomies of large metropolitan areas. But Mera finds evidence seriously lacking for this argument.
51. N. Hansen, Ibid.
52. Chenery, *et al*, *Redistribution with Growth*, Oxford University Press, 1974.
53. Ibid pp. xiv-xi.
54. I. Adelman, and C.T. Morris, *Economic Growth and Social Equity in Developing Countries*, Stanford University Press, Stanford, California, 1973.
55. Ibid, p. 186.
56. Ibid, p. 189.
57. A.G. Frank, *Capitalism and Underdevelopment in Latin America : Historical Studies of Chile and Brazil*, revised edition, New York, 1969.
58. F.H. Brookfield, *Interdependent Development*, Ibid, p. 124.
59. (*i*) J.C. Mariategui, Seven interpretative essays on Peruvian reality (transl. by M. Urquidi with a foreward by J. Basadre, from Siete Enasayos de Interpretations de la Realidid Peruana) (T) Austin, Texas, 1971; (O) Lima, 1928.

    (*ii*) C. Furtado, The Economic development of Latin America : a survey from colonial times to the Cuban revolution (transl. by S. Macedo from Lal Economia Latino-emericana : una sintesa desde la conquista Iberica basta la revoluation Cubana) (T) Cambridge, 1970; (O) Santiago, 1969.

    (*iii*) O. Sunkel, National development policy and external dependence in Latin America, *Journal of Development Studies* 6, 1969, pp. 23-48. and Transnational capitalism and national disintegration in Latin America, *Social and Economic Studies* 22, 1973, pp. 132-76.

    (*iv*) R. Stavenhagen Agrarian structure and underdevelopment in Africa and Latin America, Paper presented to an IDEP-IDS-IEDES Conference on Strategies for Economic Development : Africa versus Latin America, *Dakar*, 4-17 September, 1972. MS.

    (*v*) G.L. Beckford, Persistent Poverty : underdevelopment in plantation economies of the Third World, New York, 1972.

    (*vi*) D. Slater, "Geography and Underdevelopment" Antipode, 5, 3, 21-33. —1935, 1973 : "Underdevelopment and Spatial Inequality" in D.R. Diamond, and B. McClaughlin, (eds.) *Progress in Planning*, 4, 2, Pergamon, Oxford.

    (*vii*) P.G. Casanova, Internal colonialism and national development, in HOROWITZ, *de CASTRO* and *GERASSI*, eds. 1969, 118-39.

    (*viii*) N. Girvan, The development of dependency economics in the Caribbean and Latin America : review and comparison, *Social and Economic Studies* 22, 1973, 1 33.

(*ix*) K. Griffin, Underdevelopment in Spanish America : interpretation, London, 1969.

(*x*) Samir Amin, *Unequal Development : An essay on the social formation of Peripheral capitalism*, Monthly Review Press, New York, 1977. Also, *Accumulation on a World Scale : A critique of the theory of underdevelopment*, vols. I and II, Monthly Review Press, New York, 1974.

(*xi*) Richard Peet, Inequality and Poverty : A Marxist-Geographic Theory, *Annals of the Association of American Geographers*, Vol. 65, No. 4 (December 1975).

(*xii*) D. Harvey, *Social Justice and the City*, London, 1973.

(*xiii*) I. Anderson, Ideology in Geography : An Introduction, *Antipode*, 5, 3, 1973, 1-7.

(*xiv*) S. Folke, First Thoughts on the Geography of Imperialism, *Antipode*, 5, 3, 1973, 16-21.

(*xv*) I.G. Shivji, Class Struggles in Tanzania, University of Dar es Salaam, 1974 (mimeo).

(*xvi*) B. Chhattopadhyay and M. Raza, Regional development : Analytical framework and indicators, *Indian Journal of Regional Science*, Vol. VII, No. 1, 1975.

60. D. Slater, "Imperialism and the related spatial tendencies of capitalist development at the *periphery*". Paper prepared for the Conference on *La Cuestion Regional en America Latin* held at El Colegiode Medico, April 24-29, 1978 (mimeo).

61. D. Slater, The poverty of modern geographical inquiry, *Pacific Viewpoint*, Vol. XVI, No. 2, 1975.
Samir Amin : *Unequal Development*, Ibid.

Samir Amin states as follows about the strategy of transition for the peripheral formations : "The transition requires much more than the extension of public ownership, or the expansion of heavy industries. If it is not accompanied by a radical revision of economic choices, even if this means slowing down to some degree the maximum growth rate, such an extension contains the danger of perpetuation at the periphery the model of dependent development. The aim must be to combine the most modern instillations with immediate improvements in the poor sector in which the bulk of the population is concentrated, to use modern techniques for immediate improvement in productivity and in the situation of the masses. This immediate improvement, and this alone, can free the productive forces and men's initiative and really mobilise the whole of the people. This combination between modern techniques and immediate improvement of the situation of the masses calls for a radical change in the direction of scientific and technological research. Copying the technology of the developed world is not likely to answer the problems of the under-developed world of today. But this specific dialectic of transition does not, however, imply a rejection of the modernistic prospect."

62. J.K. Galbraith, The underdeveloped country, Massey Lectures for 1965, Fifth Series, Canadian Broadcasting Corporation, 1965.
63. Among the Social Darwinists, Malthus, Herbert Spencer, Ward and Giddings may be specially mentioned. According to Malthus, the imbalance between population growth and food supply would decree that a portion of society would be under-nourished and in poverty. Herbert Spencer applied the principle of natural selection and survival of the fittest to society, thereby reinforcing Malthus' conclusion. The poor were defined as those "naturally less fit". Ward and Giddings, by and large, adhered to the basic tenets of Social Darwinism. See Malthus, B. Thomas, *First Essay on Population*, London, 1926.
Spencer, Herbert, *The Principles of Sociology*, London, 1877. Lester F. Ward : The psychic factors of civilisation, Boston 1893. Franklin H. Giddings, *Studies in the theory of Human Society*, New York, 1922.
64. Cooley, Charles H. *Social Organisation*, New York, 1925, p. 296.
65. Robert Riendeau, "Poverty in Canada in 1964," paper presented at the Canadian Conference on Social Welfare, Hamilton, Ontario June, 1964, quoted in Harp J. and Hofley J.R. (Eds.) *Poverty in Canada*, Prentice Hall 1971; Donald R. Whyte : *Sociological Aspects of Poverty: A Conceptual Analysis*, the Canadian Review of Sociology and Anthropology Vol. 2, No. 4, 1965.
66. Peter, Townsend, (Ed.), *The concepts of poverty*, Heinemann, London, 1970.
67. D.M. Smith, *The geography of social well-being in the U.S.*, McGraw Hill, 1973.
68. Das Gupta Sugata, "Towards a no poverty society" in "Effective Anti-poverty strategies—Regional Experts Workshop organised by Friedrech Ebert-Stiftung in cooperation with ECAFE, Bangkok, December, 1973.
69. Cited by Sarva Sewa Sangh Prakashan, *Challenge of Poverty and the Gandhian Answer*. Rajghat.
70. Cited by V.K.R.V. Rao, in "Gandhian Alternative to Western Socialism"—Paper read in a Seminar on "Relevance of Gandhi to Modern Times", Delhi.

CHAPTER **4**

SPATIAL INEQUALITIES AND BACKWARD AREA DEVELOPMENT
Policies And Strategies

POLICY INSTRUMENTS BASED ON GROWTH AND DIFFUSION THEORIES

A WELTER of theories examined so far, (both growth and diffusion theories along with their extensions and modifications) contain many strands of thought and propositions that have varyingly been used for devising appropriate policy instruments in many developed and developing countries. It should, however, be noted in this context that most of the theories that we have discussed are capital-oriented theories, with concepts such as the marginal rates of savings built into them to explain the phenomenon of growth and its further transmission to other sectors and regions. An assumption common to these theories is the ultimate belief in the centrifugal mechanism of spread. They have assumed that development, whether spontaneous or induced, will start in a relatively few dynamic sectors and geographic clusters and then eventually spread to the rest of the economic and spatial systems. This will mean that if development is first concentrated in a favourable region, it will ultimately tend to benefit the lagging regions too.[1] In simple language, what is implied is : "If the rich are well off, then something will also remain for the poor". They also seem to suggest that sharp inequalities and growing unemployment are to be tolerated as the natural concomitants of an early phase of development. Hansen has appropriately labelled this group of theories as the "centre-down development paradigm".[2]

Spatial disparities in living levels have become an important policy issue both in the developed and developing countries since the 1950s and several regional policies have been devised which have essentially aimed at diffusing growth from the advanced regions to the less-developed peripheral areas. In an excellent evaluation of regional policy experiences in market and mixed economies, Walter Stohr and Franz Stodtling[3] have listed 10 major policy instruments for the development of peripheral areas. Important among them are : capital incentives and technology transfer; promotion of modern industries; transfer of public investment; extension of transport and communication networks from core regions to peripheries ; growth centre policies; extension of core-region based education and training facilities/public and private organisations/standards, rules and value systems to peripheral areas etc. Most of these policies have assumed that interregional disparities and interactions are basically conditioned by regional differences in access to production factors and to markets.

The advanced regions are supposed to be characterized by a scarcity of natural resources and of labour, while peripheral areas usually have a scarcity of capital and technology. According to neo-classical economic theory, factor returns should be high in areas of scarcity and low in areas of abundance. It, therefore, follows that the relatively developed regions will have high returns for natural resources and labour, while the less developed peripheral areas will have high returns for capital and technology. Under neo-classical assumptions (complete mobility and homogeneity of production factors, decreasing marginal returns and so forth), factors are expected to move from areas of low return to those of high returns. In the present case, natural resources and labour are expected to flow from peripheral areas to more advanced areas, while capital and technology are expected to flow in the opposite direction, that is, from the relatively developed areas to the peripheries. These flows would increase factor supply in areas of scarcity and reduce it in areas of abundance. In this manner, factor prices are expected to be equalized over space. As a result, income is also expected to be equalised over space. The neo-classical theory thus relies on the price mechanism as a spatial allocator of resources and emphasize on interregional factor mobility as a key growth determinant. From the policy angle, the inference is that regional inequalities can be satisfactorily dealt with by reducing market imperfections and by adjusting the price system via taxes and subsidies.

In reality, however, several distortions take place on account of different degrees of mobility and heterogeneity of factors and by increasing factor returns due to the unequal spatial distribution of external and scale economies. Further leakages of different types between regions (also outside the factor markets) occur which can lead to income disparities.

In the space economy of developing countries, labour and capital more often show a disequilibrating trend because migration of labour being a selective process leads to a depletion of the more skilled and more productive elements of the population strata (that is, the potential high wage earners) of peripheral areas and also causes a depressing effect on the average of wage levels. Empirical evidence also shows that on balance, capital flows from peripheral (high capital cost) areas to more advanced areas (relatively with low capital cost on account of comparative advantage in scale and external economies). The simplified assumptions[4] made in these theories as well as the growth models based on these theories[5] are, by and large, unrealistic to the Third World solution. We shall now revert to our discussion on the policy issues related to economic growth in backward areas.

In this context, we may confine ourselves to a few selected policy instruments which have been tried in many countries. It should be noted here that the treatment of the subject is in a general and broad context. The specific policies adopted in each country and the resources devoted to programmes for reducing the regional gap will depend on the nature and scale of the problems encountered, the objectives in view, the level of economic and social development and on institutional and political factors.

Among the major instruments of regional policy, one which has been used in a large number of countries is the method known as capital and technology transfer. The logic of this method is that backward areas are usually deficient in capital and technology and that the provision of these crucial inputs would render them more advanced. The philosophy of underdevelopment aid pursued by many developed countries has also followed this logic in order to assist the developing countries in catching up with the rest and closing the gap. Such capital transfers have taken a number of forms : *Concentration, stimulation and inducement.* Specifically, the measures are the following:

(a) concentrating development efforts in backward areas by locating some capital intensive or modern industries characterised by high productivity, high demand elasticity of products and/or export based orientation;

(b) stimulating the development of neo-economic activities in backward areas by making adequate provision for economic infrastructure (that is, through transfer of public investment) ; and

(c) providing various forms of public incentives or subsidies, making them area-specific that is, benefiting the relatively backward areas, (that is, inducement policies).

The effects of the above policy instruments have been evaluated in a number of studies. Generally the policy of concentration by establishing industries in selected growth areas is based on the premise that it would lead to the emergence of external economies and would result in the spin-off of development benefits from the growth areas to the entire region. However, experience shows that such developments have not occurred ; on the other hand, in many developing countries, this step has led positively to accentuation of spatial disparities in development that is, to a situation of economic colonialism. Usually this has led not only to increasing reliance on external productive factors but also to the underutilisation of the backward region's human resources. In some countries like India, it has also led to social conflicts and serious local problems. Generally the impact of this measure appears to be lost to the backward region through several forms of leakages to other regions or to elite consumption. Also it results only in small employment effects; on the other hand, its polarisation effects are great.

The direct provision of economic infrastructure mainly through investment in local infrastructure and in the interregional transport and communications network does create external economies. But such a policy, when pursued in isolation, has tended to provide only channels for leakages and has had a siphoning effect on the backward areas. Here the motivation for such provision appears to be a crucial factor. If the motivation is only to provide some linkages to external economic activities for example, the construction of a transport network which allows only goods to be exported without subserving the needs of regional development* then it serves only as a means of exploitation rather than of internal spatial integration. Experience clearly indicates that unless the provision of infrastructure is deliberately planned to subserve the needs of locally based activities, it very often produces only a negative result. Where the foreign investors are involved in setting up capital-intensive industries thereof it has also led to technological dependence and infrastructure provision and steep inequality of incomes. The resulting impact of such intervention strategy[6] is shown in Figure 1.

The provision of a variety of incentives—fiscal and monetary—in order to steer private capital into the backward areas is a measure whose efficacy has varied among different countries. While there are some instances of success, there are many instances of failures. In many South-East Asian countries, the physical planning approach to industrial promotion through construction of industrial estates has not proved to be wholly satisfactory. Lack of adequate provision of complementary infrastructure, inadequate provision of subsidy, and lack of sustained technical assistance have been pointed out as some of the weaknesses which inhibited the growth of enterprises in many backward areas. Another aspect of the inefficacy of the subsidy policy is that they have not really helped the regional inhibitants (particularly the poor and the unemployed). Historical experience suggests that the businessmen (usually from outside), the landowners and the contractors have tended to be the main beneciaries of regional subsidies. No wonder that such a strategy has not helped in narrowing interregional income disparities.

Another set of policy instruments which have been tried in many developing countries is what may be called the extension

*There are several examples of this kind in the developing countries, The D.B.K. railway line in the Dandakaranaya region serves only the limited purpose of mineral traffic from the Bastar iron ore mines, but does not serve the internal needs of passenger traffic in the area.

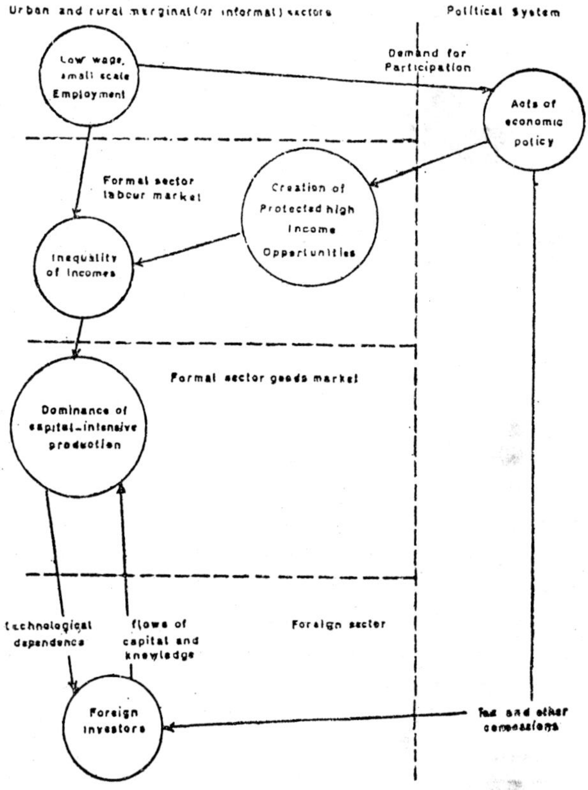

FIG. 1

strategies ; usually the base that is, the headquarter for such strategies is located in the relatively advanced regions. They have taken various forms, such as the extension of transport and communication networks from the developed regions to the peripheries, extension of education and training facilities or extension of the activities of public and private organisations whose primary operation is located in the advanced areas. In all such extension strategies, a dependency relationship is implied. They have, no doubt, led to some partial benefits ; but, by and large, they have tended to favour the concentration and centralization of efforts in the advanced areas. The extension of transport and communications network for instance has increased the accessibility interregionally, but has made comparatively little impact on intraregional accessibility within the backward areas. The extension of education and training facilities based in the advanced region has tended to increase selective outmigration from the peripheral areas, thereby increasing the disequilibrium effects. Similarly, the extension of the activities of the public and private organisations for example, banks in the peripheral areas through their "branch" operations with their managerial and decision-making functions centralized in the advanced region has often supplanted existing "all tochthonous regional organisations and thus debilitated regional organising capacity".[7]

Among the regional policies developed with the neo-classical growth theories as their base, perhaps, none excels in its extent of application the growth pole theories. They have been applied by policy-makers to serve diverse objectives.[8] As planning tools, they have been designed or instituted either to limit concentration or spread growth to lagging regions or a combination of the two. In any case, the idea is to concentrate investments in focal points so as to reap scale or agglomeration economies and diffuse development to areas immediately surrounding them. Several reviews on the implementation of growth pole strategies have appeared in recent years.[9] According to Gaile who reviewed the spread effects of growth poles in 17 different contexts, they were "smaller than expected, limited in geographic extent or less than backwash effects".

According to the empirical studies by Nichols[10] and Moseley, if the objective of regional policy is to benefit small towns and rural areas, then it would be advisable to invest directly in these places. According to them such a policy may bring about some trickle-up effect to largest cities, but not a trickle-down effect.

In empirical study by Pred,[11] the innovation diffusion process has been shown to be highly discontinuous in spatial terms ; it may be that the immediate hinterlands may not experience the spread effects, whereas distant places because of their linkages with the growth centre may experience this effect. In some Asian countries, the growth centre strategy has been used to bring about decentralised concentration of secondary and tertiary activities and to break the spatial dualism caused by primate city structures. In Japan and the Republic of Korea, the growth pole approach used as a device to accommodate industrial decentralization, has made some progress. Malaysia is also working on a similar programme in the case of Kuantan. In India, the growth centre approach has been advocated for the planning of service centres in the rural areas for the centering of minimum needs. The French planning used the growth pole strategy to lessen the economic concentration in Paris and to promote growth in other French cities outside the Paris Basin. Some African countries, for example, Tanzania, Kenya, Ghana and Gambia have adopted growth centre policies in varying degrees. In Latin America, for example, Chile, Bolivia and Colombia, the growth centre strategies have been rejected on ideological and political grounds. Stohr[12] has observed that "growth centres have essentially led to a shift of disparities from the interregional to the intraregional level, but rarely seem to have led to an overall reduction of spatial disparities in living levels".

Richardson is of the view that the failure of spread effect of growth centres does not mean that the polarisation principle is wrong. According to him, "the fine tuning needed for efficiency and balanced pole promotion may be too difficult to achieve, given the bluntness of available policy instruments".[13] He feels that the failure "reflects overambitious expectations, short-run horizons, lack of political stamina, deficient investment criteria, poor locational choices and unimaginative policy instruments".

Johnson[14] has advocated a well-articulated system of market centres for spatial development in India. Similarly, Rondinelli[15] and Ruddle have also emphasized the need to build up an articulated network of growth centres as well as linkages among them in the developing countries in order to encourage commercialisation of agriculture, savings and investment in productive activities. Hansen,[16] summing up the prevalent attitudes towards growth pole approach, refers to three phases in this development :

(1) Optimism with respect to possibilities for inducing growth in a few centres and to the subsequent generation of spread effects ;

(2) Pessimism when the expectations of the early phase failed to materialise; and

(3) a broader view of growth centre as one aspect of more comprehensive development planning.

The discussion on the growth pole strategies clearly brings out that for achieving the objective of interregional equalisation of development, the growth pole strategy per se will not be adequate (in fact, it will have more negative effects), unless such strategies are consciously and deliberately backed and sustained by complementary and supplementary measures for decentralised rural development.[17] In fact, this was one of the conclusions of a Seminar on the subject which examined the relevance of "the Growth Pole approach to Regional Planning and Development in the Asian context".[18] We shall revert to this subject again at a later stage in our discussion.

Thus the three broad groups of policy instruments—that is, capital and technology transfers, extension strategies and growth pole strategies stemming from neo-classical growth theories—which we have examined in detail clearly bring out their inadequacy to the developing countries. Stohr's analysis[19] of the ten major policy instruments identified by him in market and mixed economies also points to the overall conclusion that traditional regional policy instruments have only accentuated the existing market trends, but have not attacked the basic parameters underlying spatial inequalities in living levels. Stohr is of the view that the regional development policies have been partly successful in their quantitative aspects (numbers of jobs created, amount of investment, level of regional unemployment, regional per capita income, gross or net out-migration and so on), but in regard to qualitative and structural aspects of spatial disparities (for example, skill structure and stability of new plants and jobs, wage levels and wage differentials, organisational linkages, control mechanisms and functional characteristics of new plants and so on), these policies have either remained unsatisfactory or even deteriorated. Stohr identifies a variety of leakages, as the dominant problem associated with these policy interventions. According to him, the alternative policies might have to be oriented more specifically towards changing the basic parameters underlying the leakages of development effects out of peripheral areas.

From other angles also, the various assumptions of the strategies based on growth theories have been questioned. It is noted that the assumption that factor capital holds the key to the development process is only partially true, as it tends to lead to a patchy kind of development in the developing countries, spatially, structurally as well as socially, in the sense that it tends to bypass and marginalize sections of the community on the social matrix. These strategies have tended to ignore the critical importance of human resources which constitute one of the greatest impediments to overcoming the structural problems of economically-lagging regions. "Capital and natural resources are passive factors of production ; human beings are the active agents who accumulate capital, exploit natural resources, build social, economic and political organisations and carry forward national development".[20]

Further, a capital-oriented development has a tendency to produce goods and services which really do not meet the development requirements of the less-developed countries ; such a policy has also tended to favour the kind of imports benefiting only the higher income brackets. The unsuitability of such development to the social structures existing in the less-developed countries is quite glaring. Moreover, these policies have failed to solve the problem of unemployment and under-employment in the developing countries and have not been able to absorb the growing labour force. It has been noted that the growth theories, which overtly emphasize aggregative criteria, have delivered quantitative growth without social justice and have led to a situation in which the third world countries have fallen prey to low level equilibrium. In this context, the need for these countries to draw increasingly on the surplus resources of raw-labour and to formulate development equations in which human labour could substitute capital is being increasingly advocated. Such a shifting in development policy cannot, however, take place under the prevailing economic power constellations in the developing countries (20 per cent of the population dominating both production and distribution). In this context, redistribution with growth policies have been advocated.

RELEVANCE OF THE 'STAGES OF GROWTH' CONCEPT TO BACKWARD AREA DEVELOPMENT

The idea that development is a continuum and different stages are to be crossed before reaching the final stage seems particularly relevant to the backward area development context. This is because experience with different policies directed towards the accelerated development of backward areas in developing countries has clearly shown that a relatively long period is involved before the spatial disparities can be perceptibly narrowed down. This is because of the several constraints to be overcome, including scarcity of investible resources. Even where it will be possible to find adequate financial resources, the problem of development is not so simple, as development is the outcome of several interrelated changes, not merely in the economic sector, but in the social and political spheres as well. This suggests that the development of backward areas in the developing countries should follow a process of transition growth. Paauw and Fei[21] have advanced a transition growth hypothesis applicable to the country situations in South-East Asia. They have classified Malaysia and Thailand as examples of neo-colonial transition growth and the Philippines and Taiwan as examples of transition growth under economic nationalism.

The principles of the transition growth process enunciated by them have equal relevance to within country situations, particularly for bringing about the development of backward areas located within the developing countries. This author has followed such a line of reasoning and has extended this idea for dealing with the growth problems of backward areas in India. While this idea will be dealt with in greater detail in the chapter on India, it is necessary to set out the principles of such a hypothesis at this stage.

In the Paauw-Fei model of transition growth, the spatial structure of the economy is compartmentalized in two insulated parts : a modern export-oriented enclave and a large backward, traditional, agricultural area. According to these authors, the primary task of transition growth is to integrate these insulated parts into the national economy through the promotion of sectoral interaction. Integration contributes to the modernization of agricultural as well as to the diversification of industrial (non-agricultural) activities. In this task, the changes desired are the growth and development of indigenous factors of production (entrepreneurship, skilled labour and capital), industrialization and the modernization of traditional agriculture. How these interrelated changes could be brought about through a process of transition growth is discussed by them in their study. The related postulates in the context of a transition growth hypothesis are :

(1) In view of areal differentiation in economic, social and political developments between different areas, there will be diversity of transition experience between different areas. It is, therefore, necessary to recognize the different area typologies for devising appropriate policy instruments of transition growth. Thus patterns of transition will differ for the different typologies, the variations appearing in the economy's mode of operation and mode of organisation, thus producing different patterns of alteration.

(2) Since transition growth is an evolutionary process, the phenomenon of sequence of brief transition phases has to be recognized.

(3) The sequence of phases means that the economy is not only growing in the quantitative sense, but also that its basic rules of growth are being gradually modified. Successful transition growth requires a succession of phases in which progressively more advanced functions are acquired by the society but the emergence of each phase is conditioned by the growth accomplished in the preceding phase.

An evolutionary view of development strategy as applicable to backward areas in developing countries would mean that policies have to be continually adapted to the emergence of new transition growth phases. This would imply great strain on the planning apparatus, as, for a successful planning for transition growth, it will continually be necessary to review, monitor and evaluate past experience and chart out the future development path, based on realistic prognostications.

The study of Paauw and Fei has distinguished different sets of development strategies for the large and small labour surplus economies in the Asian region separately. In the large labour surplus countries, like China and India, employment opportunities must be primarily created by internally-oriented, large-scale, labour-intensive development works. In the case of smaller economies with small labour surplus like the Philippines and Taiwan, they feel that an alternative of producing labour-intensive manufactures for the world market, that is, export substitution is an important and feasible initial goal of transition growth.

When extending the idea of transition growth to regions, instead of nations, the differences between the two entities in regard to development planning and the fact that a region is a much more open entity than the nation must be borne in mind.

BOTTOM-UP STRATEGIES

In any case, one conclusion is compelling and irresistible from all this analysis. It is that if development in the developing countries is to be sustained, it should come from below and should be centered on the human being, the local resources and the local potentialities. This implies a reversal of the conventional process of planning from above. This is an inverted approach, planning upwards from the countryside, or from the periphery to the core. Perhaps, the Chinese experience comes very near to this proposition.[22]

In this context, a brief recapitulation of the Chinese experiment is necessary. Like many low income countries, China's economy may, at least until recently, be divided into two sectors: that which was traditional and indigenous and oriented towards supplying domestic needs, and that which was primarily foreign controlled or foreign influenced and externally oriented. The former was basically agricultural, with the village being the production base. The latter was focused on the largest city. China's policy-makers and planners have pursued three alternative development strategies since 1949 : A Soviet model of development strategy during the period 1953-57 ; the Great Leap strategy between 1958-60 ; and an agriculture first strategy since 1961.

In the first stage, major emphasis was placed on a heavy-industry-oriented pattern of development, with bias towards capital intensive methods of industrial technology. However, the agricultural expansion was sluggish. Markedly divergent growth paths of industry and agriculture contributed to increasing disproportionalities, bottlenecks and more or less severe strains in the economy, which became particularly marked by 1956 and 1957. These strains showed up in the form of tight urban food supply on the one hand and agricultural raw material shortages for industries on the other. This led to the advocacy of the Great Leap strategy which is officially defined as "simultaneous development of agriculture and industries", or "walking on two legs". This was to be achieved through the two principal means : (1) by mass mobilization of underemployed surplus labour in agriculture and (2) by much greater reliance on production processes based on technological dualism.

In this manner, the surplus labour in the countryside was not only utilised for highly labour-intensive projects in irrigation, flood control and reclamation, but also for expanding small-scale industrial production in the countryside. Thus a technological dualism was promoted during the Great Leap period with the modern sector characterised by more capital-intensive and large-scale methods of production and the traditional sector dominated by more labour intensive and small scale methods.

The Great Leap led to a great crisis characterized by acute agricultural and food shortages between 1960-1962. However, the failure of the Great Leap it could be argued was not a failure in conception, but a failure born of unrealistic expectations on the one hand, and inadequate and technically deficient implementation on the other. Based on the lesson of this disastrous experience, a new and altered development strategy came to be implemented in 1962, based on the principle that "agriculture is the foundation and industry, the leading factor", which essentially meant that planning and investment priorities were to be changed with agriculture being assigned the first place, consumer goods industries coming next, and investment goods industries relegated to a lower priority. This new approach involved a strong commitment to the technical transformation of agriculture. It also meant significant expansion in those branches of industries that support agriculture, coupled with a significant increase in the application of modern inputs to farming. Thus great attention was paid to complementary relationship between different sectors and, within industries, between different branches. *A central objective* of economic policy since 1960 or 1961, over and above agricultural development has been *self-reliance*.[23] The combined operational consequences of the agriculture first and self-reliance policies are shown in a large allocation of investment and other resources to agriculture in the form of chemical fertilisers, irrigation pumps, other types of farm equipment and rural electrification.

The shifts in development strategies in China are striking not only in regard to allocative priorities and the character of economic policies but also in the methods of implementing economic programmes. Thus, while ideological appeal—such as patriotism and working for the good of the society as a whole—was dominant during the earlier stages, they were subsequently played down and correspondingly the importance of the role of remunerative appeals (material incentives) was stepped up during the agriculture first stage. Commenting on this Alexander Eckstein observes as follows :

> In pursuing their development objectives, Chinese Communist leaders preferred to rely on normative appeals rather than material incentives. At the same time, they realised that at the present stage of Chinese development, material incentives still play an important role in motivating workers, peasants and cadres. However, it is essential to keep the operation of these incentives within well defined bounds, lest they lead to a widening of income and status differences and thus undermine socialist values.[24]

Another aspect of the Chinese model is a certain degree of decentralisation of decision-making that has been achieved in regard to planning and management of rural production. The communes (average size 4,550 households and population between 20,000 and 30,000) constitute the base level of the administrative hierarchy. But since the commune itself proved to be too large for much local decision-making its constituents (usually three levels, and sometimes four : the commune centre, the production brigade and the production team) were entrusted with increasingly local responsibility. These constituents approximate to the tradicellular structure of old Chinese society. A certain amount of decentralised decision-making has been introduced in the various levels, but at the same time keeping command over general policy.

In the Chinese model, redistribution of incomes was brought about through a series of confiscatory measures, which gradually led to the virtual elimination of private property. With the narrowing down of interpersonal inequalities in wealth and income, the regional disparities in development also appear to have been considerably narrowed down. This has been achieved through a highly centralised fiscal system with phased resource allocating powers.

It should be noted in this context that this achievement is in a certain systemic framework of a particular economic, political and social system as a whole. It is true that the Chinese experience is not all that easily transferable. But the important point is that it is in itself a fresh approach conditioned however by circumstances peculiarly Chinese. And a more important point that we may like to emphasize here is the likelihood that it may stimulate newer approaches and more effective directional departures conducive to interregional equalization of development levels.

The Chinese experiment has demonstrated that models that have struck continuity between tradition and modernity have persisted. Among other points particularly striking in the Chinese model for our present discussion on inequalities are the greater emphasis given to rural development and to the principle of equity (even at the cost of growth in output), the investment-inducement machanisms adopted to make agriculture the leading sector and the emphasis on self-reliance and the transformation of man through revolutionary zeal.

We may note these points as important directional departures or vital ingredients of any approach to development designed to reduce the regional gap within the developing countries.

A STRATEGY OF ACCELERATED RURAL DEVELOPMENT

The disillusionment with the Western growth models on the one hand and the success of the Chinese model on the other has led to a lot of new thinking for a new paradigm of development suitable to the developing countries. In the new thinking, the perception of rural development processes purely in terms of attaining techno-economic objectives has yielded place to a greater emphasis being placed on the human variable in the developmental equation and on socio-political aspects of development.

Friedmann and Douglass have recently advocated a strategy of accelerated rural development suitable to the Third World countries. This has been called agro-politan development.[25] This strategy of accelerated rural development incorporates the following policy assumptions :

Limited and specific human needs should replace unlimited, generalized wants as the fundamental criterion of successful national development

The old GNP standard, based on the idea that a sustained and rapid growth in income and production is the appropriate measure of development, would thus give way to new forms of social accounting that take as their starting point *guaranteed national minimum standards* across a wide spectrum of biological familial, and communal needs. Although gross national product accounts would be retained, they would require (1) revision to reflect the social costs of production; (2) refinement—to reflect the social valuation of specific production sectors (one dollar's worth of food not equal to one dollar's worth of automobile); and (3) expansion—to reflect the costs of production in terms of scarce energy units. Revised GNP accounts would be supplemented by social indicators showing progress made towards the attainment of specific national minima, such as protein intake and access to sanitary water supply.

Agriculture should be regarded as a leading or propulsive sector of the economy

The social efficiency of investments in agriculture is thus assumed to be high (particularly with regard to the productivity of land) and capable of generating significant multipliers in agro-industries, commerce, transportation and construction.

Attaining self-sufficiency in domestic food production should be regarded as a high priority objective

An adequate supply of domestically produced food is a basic condition for meaningful participation in the world economy. Participation would be on a highly selective basis, however, and guided by national requirements.

Existing inequalities in income and living conditions between social classes and between urban and rural areas should be reduced

The problem of distribution should not, however, be separated, either conceptually or in practice, from that of growth. Regional policy should be seen, not so much in terms of reducing disparities but of increasing generative growth through a full development of natural resources and human skills.[26] A concept of dynamic efficiency would replace the traditional criterion of allocative efficiency in capital allocation.[27]

Facilitative measures to increase production of wage goods for domestic consumption should be given high priority

This will become possible to the extent that agricultural and, more generally, rural incomes are improved. A national policy of encouraging the wage-goods sector, emphasizing small-scale production and labour-intensive methods, should help materially in the decentralization of manufacturing processes and in experimenting with new organizational settings.

A policy of planned industrial dualism should be adopted whereby small-scale production for the domestic market is protected against competition from large-scale, capital-intensive enterprises

The latter should be oriented primarily to export and/or to the production of basic materials and intermediate products. At the same time, the small-scale industry might be linked to corporate production through expanded systems of sub-contracting.

A strategy which incorporates these elements has been called by Friedmann and Douglass as *a strategy of accelerated rural development*. For the spatial framework for such strategy, a basic rural urban unit called Agropolitan District has been proposed, which is a socio-economic and political space. The essential features of such a unit are stated to be as follows :

(1) It will be a relatively small territorial unit, encompassing a rural urban population from 15000 to 150,000, depending on local population density, within a one hour commuting range by generally available transportation. The central town of this unit may have population between 10,000 and 25,000.

(2) It will have a high degree of self-sufficiency and self-reliance in decision-making and planning, based on popular participation and cooperative action at local levels.

(3) It will have diversified rural employment to include both agricultural and non-agricultural activities emphasising the growth of small-scale rural industrialisation.

(4) Its development will be based on the utilisation and evaluation of local resources and technologies.

(5) It will have both urban/rural industrial function which will have their linkages to local resources and economic structures.

The key to successful agropolitan development is the treatment of each district as a single, integrated and self-governing unit. Correlative with this idea of agropolitan development is the requirement that each unit has sufficient autonomy and economic resources to plan and carry out its own development. Taking the social structural realities into account, the agropolitan unit has to be provided in each context with an appropriate political and socio-economic system. The two important aspects of the agro-politan approach are : (a) the emphasis on local linkages and complementarities between agriculture and industry so as to internalize development, and (b) a complementary package of policies to equalize the ownership of productive assets, including land, capital and publicly created assets such as irrigation system, policy for the redistribution of income which would create demand for wage goods and thus equalise the benefits of development and promote the satisfaction of basic needs.

NOTES AND REFERENCES

1. It may be recalled that Hirschman was optimistic about the inevitability of such developments in the medium or long run; but Myrdal was more doubtful, although he was not an outright pessimist in this regard.
2. N. Hansen, Devèlopment from above : The centre-down development paradigm, (mimeographed 1972).
3. See Walter Stohr and Franz Stodtling : "An evaluation of regional policies—Experiences in market and mixed economies" in V.M. Hansen (Ed.), *Human Settlement Processes*, Ballinger Publishing Co., Cambridge (forthcoming). Also (i) Hugh D. Clout (Ed.), *Regional Development in Western Europe*, John Wiley, London 1975, (ii) OECD : The Regional Factor in Economic Development—Policies in 15 Industrialised OECD countries, Paris, 1970.
4. Some of these assumptions are: Full employment, perfect competition, one homogeneous commodity, zero transport costs, regionally identical production functions, fixed supply of labour, no technical progress, etc.
5. There are a large variety of neo-classical regional growth models ranging from naive versions to complex generalised models. According to Richardson, they have pedagogic rather than heuristic value. Some of the extended versions of these models (for example, Siebert, Friedmann) take more explicit account of relevant factors and provide better insights in the analysis of spatial problems in developing countries. These have been discussed in a later part of this chapter. For a good discussion of neo-classical regional growth models, see H.W. Richardson, *Regional and Urban Economics*, Penguin 1978, Chapter 6.
6. This representation is based on an ILO Study: *Employment, Incomes and Equality: a strategy for increasing productive employment in Kenya*, ILO, Geneva, 1972.
7. Walter Stohr and Franz Todtling: Ibid.
8. C.L. Gaile, "Notes on the concept of 'spread'," unpublished paper. Department of Geography, University of California, Los Angeles quoted in Hansen, N. : Development from above—the centre-down development paradigm 1972 (Mimeo).
9. See A. Kuklinski (Ed.), *Growth Poles and Growth Centres in Regional Planning*, Mouton, The Hague, 1972. Kuklinski and Petrella (Eds.), *Growth Poles and Regional Policies*, Mouton, The Hague, 1971.
   (i) N.M. Hansen, (Ed.), *Growth Centres in Regional Economic Development*, The Free Press, New York, 1972.

(ii) B.J.L. Berry, *Growth Centres in the American Urban System*, Ballinger, Cambridge, Massachusetts, 1973.

(iii) F. Butler, *Growth Pole Theory and Economic Development*, D.C. Heath Ltd., Hants, 1975.

(iv) M.J. Mosley, *Growth Centres in Spatial Planning*, Pergamon, Oxford, 1974. R.P. Misra, K.V. Sundaram, and V.L.S. Prakasa Rao, *Regional Development Planning in India : A New Strategy*, Vikas, Delhi, 1974.

10. F. Nichols, Growth Poles—an evaluation of propulsive effects, Environment and Planning Volume I, No. 2, 1969 pp. 193-208.
M.J., Mosley, The Impact of Growth Centres in Rural Regions, *Regional Studies*, Volume 7, 1973, pp. 80-120.
11. A. Pred, The inter-urban transmission of growth in advanced economies-empirical findings versus regional planning assumptions, International Institute for Applied System Analysis, Luxemburg, Austria, 1976.
12. Walter Stohr and Franz Todtling, Ibid.
13. H.W. Richardon, *Regional and Urban Economies*, Penguin, 1978 p. 176.
14. E.A.J. Johnson, *Organisation of Space in Developing Countr ies* Harvard University, Press, 1970.
15. D.A. Rondinelli and K. Ruddle, *Urban Functions in Rural Development : An analysis of integrated spatial development policy*, Office of Urban Development, Agency for International Development, U.S. Department of Space, 1976.
16. N. Hansen, Development from Above: The Centre-Down Development Paradigm, (mimeo), 1978.
17. The observations of Richardson in the context of Latin American development are quite significant, "The growth-pole strategy may have a place in Latin American development, provided complementary rural strategies are introduced and provided that the counter-magnets are treated only as the first phase in moulding an integrated national urban system."—H.W. Richardson, *Regional and Urban Economics*, Penguin, 1978.
18. UNCRD, "Growth Pole Strategy and Regional Development Planning in Asia", Nagoya, Japan, 1976.
19. Walter Stohr and Franz Todtling, An evaluation of regional policies —Experiences in market and mixed economies. Ibid.
20. F.H. Harbinson, *Human Resources as the Wealth of Nations*, Oxford University Press, 1973.
21. Douglas S. Paauw and John C.H. Fei, *The Transition in Open Dualistic Economies—Theory and South-east Asian experience*, New Haven & London, Yale University Press, 1973.
22. C.L. Salter, Chinese experiments in urban space—the quest for an agro-politan China, in J. Friedmann, *New concepts and technologies in third world urbanisation*, Comparative urbanisation studies, School of Architecture and Urban Planning, University of California, Los Angeles, 1974.

23. Self-reliance involves "maintaining independence, keeping the initiative in our own hands and relying on our own efforts." In the Chinese context around 1960, it carried a strong autarkic connotation, that is, emancipation from economic and military dependence on the Soviet Union. In fact, as an operational policy, self-reliance has been applied in several different contexts and for different purposes. It has been used as a means of exhortation to maximum effort and for building pride in individual accomplishments by workers and peasants, factories and commines. The desire for self-reliance is reinforced by other considerations as well, such as the urge to decentralise industrial location and to develop the more backward regions.
24. Alexander Eckstein : *China's Economic Revolution*, Cambridge University Press, Cambridge, 1977, p. 64.
25. J. Friedmann, and M. Douglass, *Agropolitan Development—Towards a new Strategy for Regional Planning in Asia*, UNCRD, Nagoya, 1976.
26. Harry W. Richardson, *Regional Growth Theory*, London: Macmillan, 1973, pp. 86-88.
27. Marvey Leibenstein, Allocative Efficiency vs. X-Efficiency, *American Economic Review*, June 1966, pp. 392-415.

PART TWO

CASE STUDIES

CHAPTER **5**

THE INEQUALITY PROBLEMS OF MALAYSIA

BACKGROUND

MALAYSIA (Area 130,000 sq. miles; population just around 12.2 million in 1975) is a nation characterised by a multi-racial, multi-religious and multi-cultural society. From the time of gaining independence on 31st August, 1957 until 1960, the nation had to endure a period of emergency during which the bulk of the planning efforts was directed towards anti-guerilla warfare. With the formation of Malaysia in 1963, determined efforts are being made towards national unity and integration. The minimization of inter-regional socio-economic development imbalances is a specific policy objective aimed at since the implementation of the Second Malaysia Plan (1971-75) and the Third Malaysia Plan (1976-80).

Politically, Malaysia is a federation of 13 States, comprising the 11 original States of Peninsular Malaya and the two Burmee States of Sabah and Sarawak which joined the federation in 1963. The political administrative system is rather centralized in terms of the power structure, as provided for in the Federal Constitution, particularly in the distribution of financial resources. The total population of the country as in 1976 was 12.23 million with 10.27 million (nearly 85%) concentrated in Peninsular Malaysia, 1.13 million in Sarawak and 0.83 million in Sabah. In terms of population density, while peninsular Malaysia had 65 inhabitants per Km^2, it was only 8 in Sabah and 7 in Sarawak. Like Indonesia, Malaysia is also blessed with adequate virgin land with potential for development.

As with many other south-east Asian countries, Malaysia's economic development too represents a transition process from colonial heritage to a modern growth epoch.

The problem in its context

During colonial times, rubber and tin were the main staples of her economy and the country's prosperity was ultimately linked with these two commodities. The resources of the country were developed through large scale immigration and large scale investment of foreign capital and the econcmy was oriented to supply the requirements of raw materials for the advanced industrial countries and to provide a market for the manufactures of those countries. Economic development concentrated along the western coast of the country, a strip of 70 to 100 miles wide running down from Penang, which is more densely populated and better developed than the rest of the country. Such a pattern of development owed its origin to two initial natural advantages along one of the great trade routes of the world for example, location and accessibility to natural resources, particularly tin and rubber. Tin mining, favoured by its occurrence along the coast and its easy extraction attracted thousands of Chinese immigrants. Later, the development of rubber plantations drew in a large number of Indian labourers[1]. During the 19th century and the first three decades of the 20th century, there was large scale immigration of the Chinese and the Indians to open up the country and develop the mines and the plantations[2]. The immigrant Chinese, Indians and others were separated from the indigenous Malays. The Chinese were involved in mining, manufacturing and commercial sectors and they lived in rubber estates and the indigenous Malays settled in the rural areas, cultivating rice, tapping rubber or catching fish.

During the colonial period, developments in Malaysia were based upon primary product export expansion. The prevalence of a free market system facilitated export orientation and the growth of an industrial sector manned predominantly for foreign enterprise. Thus the industry effectively stimulated primary export product expansion. It utilised the natural resources of the country for example, rubber, tin and palm oil and was based on the transmission of modern technology from abroad. Such development of a modern export sector led to an enclave economy which, not only contrasted sharply with the traditional subsistence agriculture, but also accentuated ethnic specialization, as it required more and more the presence of alien entrepreneurs (for example, the Chinese, the British and others). Thus a sharp dichotomy was created between the modern enclave and the traditional agricultural sector, which was further compounded by ethnic specialization.

Thus the present spatial structure of underdevelopment in the country and the population structure of the society stem from the colonial rule. In terms of spatial levels of development, a simple three-fold regional system may be recognized namely, a Central Core Region, focussing on the Klang Valley (in which Kuala Lumpur, Port Klang and a number of other towns are located), a broad resource frontier region in the interior and also in Sabah and Sarawak, and a poor or lagging region in the States of Kelantan and Trengganu on the east coast and also in an area around Penang in the north.[3]

The pluralistic nature of the society (with its cultural and religious differences) can be understood from the fact that 54 per cent of the people are Malays and other indigenous people, 35 per cent are Chinese and the rest about 11 per cent are Indians and others.

The total population as in 1976 was 12.23 million with the following ethnic breakdown :

		(millions)	%
(a) Peninsular Malaysia			
Malays (and other indigenous people)		5.49	53.5
Chinese		3.62	35.2
Indians		1.08	10.5
Others		0.08	0.8
	Sub-total	10.27	100.0
(b) Sarawak		1.13	
(c) Sabah		0.83	
	Total :	12.23	

Source : Ministry of Finance, Malaysia, Economic Report, 1976-77 Perehs-takan Mas Sdn. Bud., Kuala Lumpur, 1976, p. 5.

Thus it should be noted that the disparities problem in the Malaysian context is not merely one of spatial imbalances in development but also one associated with significant social, political and economic correlates.

The economic imbalances can be understood from the fact that in 1970, the Malays and other indigenous people owned 2.4 per cent of equity capital, the Indians held 1.1 per cent, the Chinese 27.2 per cent, others 6.0 per cent and the foreigners held 63.3 per cent.* In the distribution of income too, there exists a high degree of inequality. In 1970 in Peninsular Malaysia, households considered in poverty were about 49 per cent and accounted for 15-18 per cent of total income. The top 5 per cent of households obtained almost 30 per cent of the income. The Third Malaysia Plan document stated that, "In absolute numbers about 7,92,000 of the 1.6 million households realised incomes at or below the poverty level in 1970. The incidence of poverty was 59 per cent among rural households and 25 per cent among urban households.[4] In terms of ethnic groups, the incidence of poverty is the greatest among the Malays who are mostly found in the rural areas. The

TABLE 5.1 : **Malaysia—Poverty by States, 1975**[5]

States	Percentage of households in poverty
Johor	45.7
Kedah/Perlis	64.5
Kelantan	76.1
Malacca	44.9
Negri Sembilan	44.8
Pahang	43.2
Penang	43.7
Perak	48.6
Sabah	n.a
Sarawak	n.a
Selangor	29.2
Trengganu	68.9
Malaysia	49.3

* According to the Third Malaysia Plan document (1976-80) the targets that have been set are that by 1990, the Malays and other indigenous people will own and manage 30 per cent of the economic assets and activities, the Chinese and Indians 40 per cent and the foreigners managing the remaining 30 per cent.

Table 5.1 affords a State-wise picture of percentage of households in poverty. The States of Kelantan, Trengganu and Kedah/Perlis are characterized by very high proportion of households in poverty while Selangor has the lowest.[6]

Further insights into trends on the spatial disparities in development is afforded by a comparison of State-wise GDP per capita during 1963-70. The comparison is confined to the States in West Malaysia.

TABLE 5.2 : **State-wise GDP per capita in West Malaysia**

States	GDP per capita	Per capita GDP as ratio of West Malaysia mean		
	1970 ($)	Mean 1963-65	Mean 1966-70	1970
Selangor	1,520	1.57	1.64	1.67
Penang	939	0.98	1.02	1.03
Perak	911	1.03	1.01	1.00
Negri Sembilan	907	1.09	1.03	0.99
Pahang	855	1.00	0.95	0.94
Johor	835	0.94	0.91	0.92
Melaka	761	0.94	0.88	0.83
Kedah and Perlis	605	0.69	0.68	0.66
Trengganu	536	0.63	0.59	0.59
Kelantan	420	0.49	0.44	0.46
West Malaysia Mean	912	1.0	1.0	1.0

Source : Economic Planning Unit, 1974.

From the above Table 5.2, it may be seen that Selangor, the richest State enjoyed a per capita income of 1.57, 1.64 and 1.67 times the national average respectively during 1963-65, 1966-70 and 1970, while Kelantan, at the bottom of the scale, had a per capita income of less than half the national average. While the per capita income of the richest State increased over the years, that of the poorest State declined during the same period. The range between these two States was about 3.2:1 during 1963-65 and 3.63:1 in 1970 that is, the regional disparities increased.

To sum up, the development problem in Malaysia is characterized by (a) marked spatial variations in development, (b) economic dualism, characterized by a modern plantation and mining sector (manned by the non-Malays) co-existing with a traditional subsistence sector (which is almost completely Malay), and (c) ethnic pluralism which is characterized by races who are identified with their specialized economic functions and are at different levels of development.

Plans, strategies, policies and priorities

Since achieving independence in 1957, Malaysia has been tackiing the above categories of imbalances through a series of 5-year plans : the First Five Year Plan (1956-60); the Second Malay Plan (1961-65); the First Malaysia Plan (1966-70); the Second Malaysia Plan (1971-75) and the Third Malaysia Plan (1976-80). One of the basic differences in the approach to the various plans is that prior to 1970, they focussed mainly on economic growth alone and very little on income distribution. On the other hand, the Five Year Plans commencing from 1971 are emphasizing on both economic growth as well as equitable redistribution and restructuring of assets.

Thus the primary task of national development policy since 1971 is being guided by the objective of minimising inter-racial and inter-regional socio-economic development imbalances. This is to be achieved through the New Economic Policy (NEP) formulated during 1969-71 which has two objectives, (1) to eradicate poverty among all Malaysians and (2) to restructure Malaysian society so that the identification of race with economic function and geographical location is reduced and eventually eliminated. Both objectives are in many respects interdependent and mutually reinforcing. The NEP is being implemented through the strategies and policies outlined in the Second and Third Malaysian Five Year Plans.

The first objective of the NEP aims at improving the economic condition and equality of life of the poor of all races, by increasing their access to land, capital, and other public facilities and their opportunity for education and training. The second objective aims at a more equitable distribution of income, assets[7] and other opportunities in the various economic activities of the country.

An important aspect of the strategy of development is that designed to bridge the gap between the modern enclave type of development with the traditional agricultural sector. This was a problem which the Government of Malaysia had to tackle soon after independence. It inevitably demanded a rural focus.[8] Accordingly, the newly formed independent government initiated a transition process in the rural areas by seeking to develop the traditional agricultural sector by diverting a part of the agriculture surplus to rural infrastructure development (for example, roads, land development, irrigation, electrification works). One estimate places the expenditure incurred in rural infrastructure programmes during the period 1963-65 at 59 per cent of total government expenditure, equivalent to at least 1/8th of the economy's real production.[9] Hand in hand with these rural infrastructure programmes, efforts were concentrated to improve agricultural extension, education and public health services.

The long-term strategy of rural development in Malaysia is characterized by a four-step progression as follows :

(i) Phase I—Emphasis on 'providing amenities' and building physical infrastructure;

(ii) Phase II—Emphasis on people taking full part in the development process, with emphasis on Kampong development communities, adult education and community self-help activity;

(iii) Phase III—Emphasis on rural marketing and credit;

(iv) Phase IV—Emphasis on the reduction of rural urban imbalances with more rural industry and enterprise.

Phase I was launched in 1960, Phase II in September, 1962; Phase III was nominally launched in 1964.

In regard to the implementation of rural development programmes, Malaysia is well known for its Operation Room and Red Book techniques. Although these were basic instruments developed during the emergency period (1956-60) in the war against the Communist guerillas, they have continued to be applied in the area of development planning and implementation with a number of modifications. Essentially the operations room consists of the basic data (maps, projections, charts) which facilitate monitoring of the progress of programmes and identification of bottlenecks. The Red Book system provides the guidelines and the basic framework for development planning and method of implementation of development projects.

The above inherited system of rural development is undergoing reorganization with a view to developing a new development planning system consistent with the changed and much more ambitious national goals and objectives. A number of new institutions have been established such as the Economic Planning Unit and the Implementation and Coordination Unit at the Federal level and the State Development Committees at the State level.

In the Malaysian situation, the development of the traditional agriculture sector and reduction of the ethnic imbalances in the country are inevitably linked to a backward area development strategy which is essentially a decentralization strategy aimed at the transfer of surplus resources from the developed regions, for example, the Klang Valley to the rest of the country. This decentralization strategy is sought to be synchronized with land development on the frontier zone in order to shift some of the population as well as to focus on agricultural and rural development programmes. The whole series of regional development plans which have been prepared for the areas located in the lagging regions in the east and the interior such as the Pahang Tenggara, Trengganu Tenggara and Kuantan are intended to bring about a spatial reorganization of the Malaysian economy by linking land settlement with exploitation of the natural resources. There is also the Klang Valley Plan which seems to rationalize spatial industrial planning in the Klang Valley. Through these various plans, Malaysia is proposing to implement a strategy of dispersing industries to the less developed regions from the Kuala Lumpur-Klang Valley area. The fiscal policies in the country have also been oriented towards the above strategy. Notable among the fiscal policies are the various investment incentives provided to the less developed areas in order to promote dispersal of industries. Factories located in areas declared development areas are granted one additional year of tax relief, if it is a pioneer company, or an extra 5 per cent tax credit, if it is company granted investment credit. In addition, a new set of incentives called locational incentives have also been instituted, granting exemptions varying from 5 to a maximum of 10 years, depending on the size of capital expenditure and employment of the enterprise. The locational incentives areas apply mostly to the less developed parts of the country. The Federal Industrial Development Authority (FIDA) as well as other State Government agencies are engaged in various tasks of promoting and coordinating industrial development which include the actual setting up of industries or taking part in joint ventures or establishing industrial estates. Thus in Malaysia, the dispersal of industries and regional development are becoming powerful instruments for developing the backward regions. As part of the industrialization strategy, the government is also vigorously promoting, through vocational training and education, the improvement of the skills of the indigenous population in order to build them up into entrepreneurs.

Land development, infrastructure development and regional development schemes have undoubtedly spearheaded the backward area development strategy in Malaysia. They are truly exercises in social engineering as well as experimentation with organizational and institutional innovations. They contain success stories with valuable lessons in methods of promoting the planned flow of labour resources from congested production centres to new growth poles in the agricultural sector. Perhaps some elaboration in this context may be in order.

Malaysia exhibits various types of land development schemes carried out by different agencies.* They differ from one another in the extent of the degree and scale of financing and investment, the amenities provided for, the scale of operations and the role of the participants. Among these, the FELDA (Federal Land Development Authority) model of development deserves special mention.[10] Land development, as carried out by FELDA, utilizes the package deal approach in which it is engaged in the various operations and activities through

* Among them are: the Federal Land Consolidation and Rehabilitation Authority (FELCRA), the Rubber Industry Small holders' Development Authority (RISDA), the State Land Development agencies and the Federal Land Development Authority (FELDA).

land clearing, planting, settlement, the provision of credit, processing and marketing facilities. The FELDA land development programme is informed by economic as well as social objectives. Increasing the income level through the utilization and exploitation of land for an efficient agricultural production is the prime economic objective of land development. The realization and achievement of the economic objective through the settlers' own efforts and resourcefulness and the participation and involvement of the settlers in the development process is the important social objective. Land development and settlement involves the planned and controlled transfer of people from one area to another, which implies problems of adjustment and adaptation to a new form of living. From this standpoint, the FELDA scheme is an important attempt in social engineering. Judged by various criteria, it is stated that the FELDA model of land development is a success. This model has a built-in mechanism to produce, identify, act upon and resolve the various problem areas that emerge through the development process. Its resilience and flexibility which have continued to improve and build up FELDA's policy and operations, are demonstrated in the changes and modifications related to the size of holding, settlement patterns and tenure systems.[11]

In the Malaysian situation, the FELDA model appears to be conducive to *equal* development, since it seeks to give every farm family equal access to enough agricultural land to enable them to produce enough to rise above the poverty line. This conclusion is borne out from a study on the distributional impact of the green revolution in regions of Malaysia and Indonesia carried out by David Gibbons and others.[12]

Apart from the FELDA type of programmes, certain regional infrastructure building programmes implemented in Malaysia have also afforded valuable insights into the factors that influence the change process to bring about distribution changes in income and wealth. The Muda Irrigation Project on the north west coast of Malaysia may be cited as a project engendering some valuable lessons in this context.[13] The factors that contributed to its success in integrated rural development planning are: (a) the vesting of the administration and management of the project to an executive authority as the sole agency responsible for socio-economic development of the region; (b) giving equal emphasis not only to physical construction works but also to the agricultural service component and the human resource development component; and (c) ensuring grass roots level participation through appropriate institutional mechanisms like farmers' associations and work groups.

A third category of regional development plans may also be mentioned, which have visualized a model of development, both agricultural, industrial and urban development in areas formerly under jungle. A number of them could be listed,[14] but excepting the Pahang Tenggara, they are all of limited scope and do not realistically fit into the context of development of the Malaysian economy as a whole.

The Pahang Tenggara is an improvement upon the FELDA type land settlement scheme. It links land settlement and national resource exploitation along with an urban development strategy, geared to not merely the regional problem but to that of the nation as a whole. Its urban system design creates a regional hierarchy which has been sought to be integrated into the national system of cities. It combines a migration strategy to draw Malay labour from less productivity farms into the high productivity activities centered in its planned cities and the surrounding hinterland. Among the objectives of the project is the "promotion of substantial amounts of industrialization and urbanisation of a kind which provides opportunities for Malays as well as other Malaysians, to find employment at all levels in the modern sector, as well as to contribute to the reduction of regional disparities."[15]

To sum up, in the case of Malaysia, regional and racial equity policy objectives are embodied in the New Economic Policy of the government. The government has set itself the time period of twelve years that is up to 1990 to achieve these objectives. In order to achieve its objective of restructuring society, the government have set up public enterprises* with the principal aim of injecting *bumiputra* (sons of the soil) participation in all sectors

* Such public enterprises are : PERNAS (the National Corporation), SEDCs (State Economic Development Corporations), UDA (Urban Development Authority), MIDF (Malaysian Industrial Development Finance) and MARA (The Council of Trust for the Indigenous People).

of the modern economy. Given the official policy towards foreign capital in the past and the low rate of capital accumulation among the Malays, it is difficult to see how the Malay community can muster sufficient capital within the next decade to achieve the target of 30 per cent ownership of the private sector. Similarly given the very limited pool of Malay managerial talents, there is perhaps even greater difficulty in the short time in trying to meet the targeted level of Malay management. It must be emphasized in this context that the correction of racial inequality can be undertaken in the context of rapid structural change and expansion of the economy. The *bumiputra* policy, when carried to the extreme will result in a sense of deprivation among other races and this may tend to slow down the growth rate of the economy. Such dangers are inherent in the present strategy.

The Third Malaysia Plan has, as its basic thrust, a multidimensional approach in tackling poverty in the less developed areas. It is stated that "the main focus will be on the bottom 40 per cent of the households in the distribution of income, which account for only 11.5 per cent of the total income in the country." The tactics adopted to reduce poverty are to raise productivity in agricultural areas and remove agricultural labour in low productivity occupations to higher productivity activities like manufacturing. To implement the rural development programmes, a multiplicity of institutions have been set up aiming at servicing the peasantry by providing extension, credit, and marketing services. Through bold land settlement and regional development programmes and locational incentives, a dent has been made into the problem of uplifting backward areas and promoting balanced regional development. For obvious reasons, such development calls for a continuing and long term effort. There still exists stark regional economic disparities between the backward states of the east coast and northern peninsular Malaysia and the west coast states of peninsular Malaysia. Notwithstanding these limited achievements, the development experience of Malaysia must be deemed more successful among most developing countries.

NOTES AND REFERENCES

1. Malaya came under British control gradually between 1786 and 1895. Tin mining had been carried on by the Chinese even before the advent of the British, but rubber was first introduced only in 1877.
2. The Chinese immigration took place earlier and at a faster rate than Indian immigration.
3. Kamal Salih, Industrialisation, strategy, regional development and the growth centre approach—A case study of West Malaysia in *Growth Pole Strategy and Regional Development Planning in Asia*, United Nations Centre for Regional Development, Nagoya, Japan, 1976.
4. Government of Malaysia, *Third Malaysia Plan 1976-80*—Government Press, Kuala Lumpur 1976, p. 5.
5. Third Malaysia Plan, p. 5.
6. Third Malaysia Plan, p. 202.
7. Among the targets that the Third Malaysia Plan have set are that by 1990, the Malaysian citizens will own and manage 70 per cent of the economic assets and activities of Malaysia, with the foreigners managing the remaining 30 per cent. Among the Malaysian citizens, the Malays and other indigenous people will own and manage 30 per cent of economic assets and activities, the Chinese and Indians 40 per cent.
8. The rural focus is also a political necessity, as with it goes the improvement of the welfare of the Malay component in the population.
9. Golay, *et al*, *Underdevelopment and Economic Nationalisation in South-East Asia*, Cornell University Press, Ithaca, 1969.
10. For a good account of Malaysia's land development programme as carried out by FELDA, see Alladin Mashim, Land development under FELDA : Some socio-economic aspects in *Rural Development in South-East Asia*, Ed. A.R. Mokhzani, Vikas Publishing House Pvt. Ltd., New Delhi, 1979.
11. Alladin Hashim, Ibid.
12. David S. Gibbons *et al*, *Agricultural Modernization, Poverty and Inequality*, published by Teakfield Limited 1980. According to this study, "In predominantly agrarian societies following a 'Western' model of economic growth, substantial and rapid reduction of poverty and more equal development probably require (in fact if not also in theory) more equal access to agricultural land as well as modernization of investment for the greater share of public sector investment for the rural areas". The region selected for the study in Malaysia comprises the North-Western States (Perlis, Kedah and Penang) centering on the Kedah plain and on the Muda scheme, the rice bowl of Malaysia.
13. For a case study of Muda Irrigation Project, see the paper by L.J. Fredericks, included in United Nations ESCAP : *The Practice of Local-Level Planning—Case studies in selected rural areas in India, Nepal and Malaysia*, 1980, Bangkok.
14. A partial listing will include the plans for Trengganu, Penang, Johore Tenggara, the Klang Valley, South Johore, Kelantan and Pahang

Tenggara. They are plans for small regions which are neither economic regions nor political units. Benjamin Higgins has made a comparative analysis of these plans. According to him "except for Penang, and to some degree for Trengganu, the earlier regional plans are really land settlement schemes. They reflect an earlier vision of regional planning which involved the successive clearing and populating of empty under-populated areas, so as to expand the export base of the national economy. With the study of Pahang Tenggara, this phase of regional planning came to an end." Benjamin Higgins : A Comparative Analysis of Regional Plans in Malaysia (Mimeo).

15. Pahang Tenggara Master Planning Study—The Master Plan 8.1.

CHAPTFR 6

THE REGIONAL PROBLEMS OF THE PHILIPPINES

THE PHILIPPINES (population around 37 million in 1970 growing at 2.8 per cent per annum) is an archipelago of innumerable small islands, numbering about 7055, a large number of which are still uninhabited. Eleven of the large islands constitute 92.3 per cent of the total land area of the country. The population of the country is unevenly distributed with more than three-fourths concentrated in the islands of Luzon and Visayas.[1] The island of Mindanao is somewhat unique. Although supporting only 22 per cent of the total population on 34.25 per cent of the land surface of the country, it is experiencing a population growth rate of 4.57 per cent which is considerably in excess of the national average. Provincial population densities range from approximately 16 per sq. km. (Palawaan) to 423.6 per sq. km. (Pampaanga) and the maximum being 1388.5 per sq. km. in urban Rizal. 85 per cent of the people of Philippines are Roman Catholics. Islam accounts for 4.2 per cent and the rest constitutes other religions. The age structure of the population with about 68 per cent in the age group below 24 years shows a high dependency ratio.

Like that of many underdeveloped countries, the economy of the Philippines is characterised by primary product export orientation, comprising a large sector producing raw materials and other primary goods* and a backward peasant agriculture.

* These consist of agricultural, mineral, forest and fishery resources. Among agricultural resources, rice, corn, sugar, coconut, abaca, tobacco, and banana are important. The mineral resources include primary gold, nickel, copper and various other metallic and non-metallic resources. (*Contd.*)

Although manufacturing involving substantial local processing of sugar and coconut products has developed, the economy of the Philippines is primarily non-industrial.

FIG. 3

Like most countries in south-east Asia, the Philippines has had a long history of colonization—more than 300 years of Spanish rule from 1521 to 1896, followed by American occupation from 1896 to 1946. Independence came in the wake of

About 30 per cent of the land area is under forest. There are 2400 known fish species in its territorial waters spreading over 180 million hectares and about 120 fishing grounds. During 1975, the primary sector (agriculture, forestry and fishing) employed more than half the labour force. Sugar is the leading export crop, (with 25 per cent export earnings) followed by copra and coconut oil. Most of the export items are subject to world market price fluctuations.

post-war rehabilitation, economic scarcity and social dislocation. The foreign exchange reserves which had accrued to the country at the time of independence were seen wasted on non-essential imports. By 1949, a serious balance of payments situation had arisen. The government responded to the crisis by clamping exchange controls and initiating economic and social reforms. Essentially the thrust was on import substitution. A vigorous industrial expansion—a whole range of domestic industry expansion—took place between 1952-56, when foreign exchange earnings from traditional exports were redirected to serve the needs of the manufacturing sector. The policies which guided this development may be grouped under three major categories : (a) tax incentives policies; (b) tariff policy; and (c) exchange and import control.[2] The cumulative effect of these policies was a heavy concentration of industrial establishments in Metropolitan Manila, contributing to accentuated polarized development. In spite of the corrective measures introduced by the Board of Industry to include within its industrial priorities formula, a factor to assist industries located outside Metropolitan Manila obtain a favourable treatment, no significant changes resulted.

TABLE 6.1 : **Regional income estimates, Philippines, 1974**

	GDP per capita (1974)	% of variation from National GDP
I. Ilocos	1431	(—) 40.8
II. Cagayan Valley	1325	(—) 45.2
III. Central Luzon	1780	(—) 26.3
IV. Southern Tagalog	4421	(+) 82.8
V. Bicol	1187	(—) 51.0
VI. Western Visayas	2677	(+) 10.7
VII. Central Visayas	2016	(—) 16.6
VIII. Eastern Visayas	1304	(—) 46.1
IX. Western Mindanao	1472	(—) 39.1
X. Northern Mindanao	1775	(–) 26.6
XI. Southern Mindanao	2755	(+) 13.9
XII. Central Mindanao	1426	(—) 41.1
PHILIPPINES	2418	

Source : NEDA

The above account is intended to provide a background to the regional problems of the Philippines. These are reflected in the regional income gaps in the country which are brought out by Table 6.1.

It may be seen from Table 6.1 that Region IV which includes Metropolitan Manila dominates the rest of the regions with a gross regional income accounting to about 47 per cent of the gross national product. Three regions of the country, Eastern Visayas, Bicol and Cagayan Valley constitute the relatively backward regions of the country with per capita gross regional incomes less than half of the national average.

In the scale of poverty, the pattern is similar to the other countries studied in this volume. The absolute number of people in poverty is estimated at 14 million (1975) which is about 33 per cent of the population. The share in GNP of the lowest 40 per cent of the population is only 11.6 per cent.* There is evidence that the proportion of poor families is increasing.** Unemployment does exist, but more important than this is underemployment in the sense of not having enough work to do, as well as not earning enough to support oneself and family.[3]

We may examine the policies, programmes and plans designed to decrease the inequality of incomes in terms of areas/regions and families against the socio-economic backdrop provided above. A reference was made earlier to the industrial promotion policies pursued in the fifties. They were not specifically aimed at bringing about regional dispersal of industries.

* See Tables 1 and 2 in "Growth and Poverty in Developing Countries", World Bank Staff Working Paper No. 309 (Revised), May 1979.

** A study conducted by the Development Academy of Philippines concluded that in 1971, 56.9 per cent of families were below the total threshold i.e. below the barest minimum by which subsistence can be theoretically achieved. There are no figures beyond 1971 for family income. However, data on the consumer price index are available. Unless family incomes go up in proportion to the consumer price index, the number of poor families are expected to increase. The trends seem to indicate that wages are rising slower than the consumer price index. It may be therefore inferred that the proportion of poor families is increasing. Proserpina Dominge-Tapales : Planning for Implementation, the Philippines case, Paper contributed to Workshop on Improving the methods of Planning for Comprehensive Regional Development, 16 May—12 June 1978, United Nations Centre for Regional Development, Nagoya.

In the circumstances, they only reasserted and expanded Manila's dominance in the economy. From the fifties to the seventies, the performance of the economy, in terms of balance of payments and the foreign exchange position on the one hand and of the sector producing for domestic consumption on the other, has led to a distributional pattern which has left the poorest groups largely outside the sphere of economic expansion and material improvements and has also accentuated the regional imbalances in the country. After the declaration of Martial Law by President Marcos in 1972, several changes have followed. The central planning agency—the National Economic and Development Authority (NEDA) was formed which came up with the Four Year Development Plan for 1974-77. This has been followed by a detailed Five Year Plan for 1978-82 along with a long-range plan for the year 2000. It is interesting to observe that the Four Year Development Plan (1974-77)* had, among others, two specific objectives of achieving (a) more equitable distribution of income and wealth, and (b) regional development and industrialization. These were sought to be achieved through programmes of integrated regional projects to redress the glaring imbalances among the regions, land reform programmes, development of cooperatives and other programmes. Similarly one of the objectives in the 1978-82 Plan is "increased development of lagging regions, especially rural areas". The Plan for 1978-82 calls for, among others, diversification of the industrial sector, expansion and diversification of agricultural exports and self-sufficiency in food, human resources development and education, and greater participation of local governments in the developmental efforts. Thus the concern for equity, for promoting planning at the sub-national levels and for rural development through local level planning are in evidence in the programmes aiming at the development of the New Society.

* The 1974-77 Plan set forth the following objectives: (1) maximum utilisation of the labour force through the promotion of employment, (2) maximum economic growth feasible, (3) more equitable distribution of income and wealth, (4) regional development and industrialisation, (5) promotion of social development, and (6) maintenance of an acceptable level of price and balance of payments stability.

In order to give effect to the above strategy, the country was divided into 11 administrative regions (later 13).[4] Each administrative region has a designated regional capital which will serve as the key growth point of the region. In each region, a Regional Development Council (RDC)* which takes the primary responsibility for planning and formulation of policies for that region, has been established. The regionalisation move is an attempt to impart efficiency in administration through deconcentration of administrative powers to regional offices of national agencies, and through easier coordination by the location of offices in the same regional capitals. Regional budgeting has been adopted which provides the field offices with their own resources for the planning function.[5] The RDC is essentially a planning advisory body with neither implementing powers nor control over the regional budget. It derives its technical support from NEDA with the Regional Executive Director (RED) performing a vital role. Plan formulation is done through task forces composed of the members of the RDC through their planning offices and local development staffs. Plans formulated in the regions are reviewed by the NEDA's Regional Development staff at the head office.

The above type of regionalization with the designation of a regional capital in each region must be seen as an attempt to break the dominance of Metropolitan Manila and as a move towards multi-polar development.[6] All the regions in the Philippines have organised their respective RDC's. Some have already submitted their respective Regional Development Plans to the NEDA.[7] The NEDA has provided some broad guidelines embodying the national policy with respect to the development of the regions. To what extent, the regional plans of the RDC's have become powerful instruments to influence the national plan is a question which is difficult to answer at the moment. Considering that RDC's have no implementing powers and financial base, probably they generally tend to toe the line of

* The RDC is composed of the Governors of all provinces and the mayors of all cities within the region and the Regional Directors of development planning and administrative organs of the Government in the area. Although chairmanship of the RDC was performed by one of the local chief executives elected by the members, the position is now filled by appointment by the President.

the national plans and build their own programmes and projects around the national plans and programmes. However, the fact remains that the RDCs exert their own pressures and try to maximise their share of the national programmes. Such pressures stem out of a high degree of interest of local elective officials and people in the development of their region. The two problems facing the RDCs are : coordination and planning capacity. The RDC is struggling for an elusive unity of actions, unarmed with sanction to elicit or enforce cooperation from the individual departments.[8] Closely linked to the problem of coordination are the problems of leadership of the Council and the general issue of representation in the Council itself. Presently the Chairman of the Council is a Presidential appointee (not an elected chairman by the RDC members) and the structure of the RDC is inadequate to provide for active participation of the citizenry and the private sector. The other problem referred to namely, the planning capacity of the RDCs, stems from the inadequacy of regional planners, incapacity of local governments and insufficiency of funds.

Undoubtedly the RDCs are still struggling to find their feet in a dynamic situation. This is apparent from the emergence of another parallel attempt at area planning and coordinated implementation of rural development projects called Integrated Area Development Program (IAD) described in a later context in this chapter. Presently, this latter programme does not fall into the RDC's orbit and is operating directly under a Cabinet Coordinating Committee. Institutionally, the two overlap (a jurisdictional dilemma). Philippines planners consider this, however, as a healthy experiment,[9] as several administration models are being tested.

Before we close our discussion on the regional development strategy in the Philippines, it is necessary to note that the Philippines has also some experience in frontier settlement projects, although in the policies and programmes of the Philippines government, there are very few measures for coping with internal migration directly. In the early 1900s, the movement of people to agricultural settlements in the Cagayan valley, Mindoro and Mindanao was a response to land hunger as well as conscious governmental policies to encourage growth in these areas. Since 1939, the Southern Mindanao provinces of

Catabato, Davaao and Zamboanga have consistently shown positive balances of net-in-migration. This type of internal migration attracted by vacant agricultural land has almost dwindled since the seventies. It is in this context that the importance of regional development programmes to realise the regional potentialities and to encourage people to remain where they are must be openly acknowledged. Otherwise, there is a grave danger that the rural urban stream toward the Metropolitan Manila area may accelerate.

Another prong of the Philippine strategy to achieve a more even development is through rural development. This rural development strategy has several ingredients of policy and strategy, the chief among them being (a) policies to improve the assets position of tenant farmers and landless tenants; (b) provision of much greater direct support to productive activities, upon which the poor are heavily dependent; and (c) selective integrated area development strategy.

Agrarian reform constitutes the basic strategy in the movement towards the New Society. A significant event in the immediate post-1972 period was the declaration of all rice and corn areas as land reform areas, with the tenants of those lands being deemed owners of the land they tilled as at the time of the decree in 1972. The setting up of farmers' cooperatives to assist the new leasehold tenants, left without any alternative source of credit to that formerly provided by the landlords themselves, is an important adjunct programme. This was delayed and other factors* also hampered the pace of progress of the land reform programme. It needs no emphasis that without a comprehensive range of innovations, agrarian reforms cannot make their desired impact in a short period.

Among government sponsored food production programmes, implemented through an aggressive extension service, using a

* Other factors besides delays in setting up farmers' cooperatives include (a) the complex procedures of the courts of agrarian reform and the inadequacy of survey teams and of legal and administrative personnel, (b) the malpractices of the landlords including the eviction of tenants and conversion of rice and corn land into other uses so as to remove it from the ambit of the land reform decree, and (c) the provision in the reforms enabling landlords to retain 7 hectares each if they become owner cultivators. Since about half the landlords exercised this option, the potential land area for distribution was inadequate.

package of innovative technology including credit must be mentioned the two crash programmes—the Masagana 99 for rice production and the Masaganany Maisan for corn production. The success of these programmes remains uncertain.

Realizing that the past efforts in rural development had been fragmentary and uncoordinated, the government have embarked on an integrated approach to rural development. For this purpose, a Cabinet Coordinating Committee for Integrated Rural Development Projects (CCC-IRDP) was established in 1973 to select priority areas for integrated development, prepare project studies, negotiate funding and supervise implementation.[10] The main criteria for selecting priority areas were their relative status with regard to underemployment and their potential for rapid development. Thus this development approach is based upon the premise that development efforts targeted on the rural sector and the small farmer should focus on delimited geographic areas of high growth potentials. Presently the programme has been launched in four priority areas namely, Bicol, Mindoro, Cagayan Valley and Samar. The integrated projects formulated under this programme seek to redress unemployment and low labour productivity, increasing regional and local imbalances, inadequate and inefficient delivery of essential services and the misuse of indigenous resources. The CCC-IRDP, which is the organization charged with this integrated rural development project, is composed of Cabinet Ministers whose ministries are directly involved in the delivery of services to the rural areas. In order to ensure the coordination of this programme with the total national development plans, the CCC-IRDP is placed within the framework of and under one administrative supervision of the National Economic and Development Authority (NEDA), the country's central planning agency.

It is perhaps useful to examine the problems encountered during the implementation of this programme, as they may have some useful lessons. Three sets of problems may be mentioned :

(*i*) the problem of coordinating the multifarious activities and projects cutting across several sectors and agencies;

(*ii*) the problem of the scarcity of resources : Capital and more critically of personnel resources; and

(*iii*) the problem of evolving a machinery appropriate for the effective planning, execution and monitoring of integrated rural development projects.

The first problem[11] namely, coordination with numerous departments and agencies has been recognised as the most difficult one. In the Philippines, this problem has been tackled in a number of ways such as through the creation of interagency task forces or committees and through instruments such as the memoranda of agreement.

The second problem namely, paucity of human resources equipped with the technology and skills and properly motivated for the various tasks, is sought to be tackled through a number of training programmes. This is, however, a slow process and there seems to be no alternative to this problem other than going ahead with the implementation of the project along with on-the-job training activities.

The third problem hinges on the organizational mechanism for planning and implementation. Should it be a strong regional or project authority coordinating all other agencies in the project areas during all the stages of the projects development? Or should it be an inter-agency council or committee closely coordinating the planning activities of the different agencies involved in one way or the other with integrated rural development projects, but leaving their implementation essentially to the agencies concerned? Irrespective of the kind of mechanism adopted, the Philippine experience indicates that it is the presence or absence of strong grass-roots organization (for example, irrigators' service associations[12], fishermen's cooperatives) that would spell the difference between a truly successful integrated rural development project and one that is not. This provides a clear demonstration of the importance of participative approaches in local-level planning.

The IAD approach is considered to be effective, as it is pragmatic, problem-oriented and dynamically flexible. It seeks to unsnarl development bottlenecks and fill in strategic gaps. During the working of the IADs, several problems have been faced. One of the lessons which has become clear is the fact that drawing up of a good plan alone is not sufficient; it is in implementation that the greatest obstacles have to be faced, and these are mainly administrative in nature.[13] How some of these problems have been solved in some of the IADs provides valuable lessons. The Bicol River Basin Council, one of the successful IADs in the Philippines created an inter-agency study team to coordinate sectoral implementation of programmes. It also obtained powers for budget review for all projects within its geographical jurisdiction. It also introduced a management information system, somewhat on the lines of the Malaysian Operations Room Technique to strengthen performance evaluation and review. According to Santiago, the Philippine IAD experiment is a pragmatic venture. "Its form secedes from the elegance of comprehensive planning and goes beyond an incrementalist strategy in its drive for a handle on the implementation.[14]

It should be noted that the integrated approach to rural development as followed in the Philippines is a very selective programme confined to certain priority areas. As regards the non-priority areas, they have been left to the other government ministries and agencies in rural development. Coordination of their activities through the thirteen Regional Development Councils is being attempted. Since however, they do not have the full authority and responsibility in decision-making, this is only partially effective.

Besides the regional and rural development efforts mentioned above, efforts are also being made to strengthen grass roots government and local-level participation in the development process, although it must be admitted that in the Philippines, decentralised development planning in the true sense of the term is still far from reality. Local autonomy measures include the nurturing of the traditional village organisations called Barangay councils and assemblies. Through these, residents in the villages have indirect access to policy-making within their areas. But more than in the field of plan/policy formulation, the Barangay network seems to have been efficient instruments in implementation of programmes, in getting people to cooperate in developmental activities.

To sum up, a grand social and economic transition seeking to change as quickly as possible from a long-standing system of subsistence agriculture and primary product export to diversified production and fluidity in monetary matters accompanied by a desire for more even development is in progress in the Philippines. This is not a simple matter. Regionalization of the country and measures undertaken for rural development have laid the basis for new activities to redistribute wealth and to achieve a more balanced development. These strategies represent a distinct departure from the policies of the 1960s. In the face of a continued high growth rate of population and in the context of opposing forces at work that do not permit the filtration of the distribution of growth benefits to the less privileged, it is difficult to foresee much improvement in the distribution of incomes in the rural areas. Spatial shifts in development and decentralization of economic growth opportunities are, of course, closely linked with the possible changes in the private sector. In the long run, the regionalization scheme, if effectively implemented, may lead to a more deconcentrated type of development by harnessing the extreme sense of regional identity of the Philippines for development purposes. Basic to this outlook are the changes necessary to strengthen the ability of the economy to earn foreign exchange and reach towards a more viable balance of payments position, while simultaneously providing additional work opportunities for the growing labour force.

NOTES AND REFERENCES

1. The concentration in Luzon is 52.2 per cent and in the Visayas Islands, it is 25.85 per cent.
2. For a discussion of these policies, see H. Power John and P. Gerardo Sicat. *The Philippines : Industrialisation and Trade Policies*, Oxford University Press 1971; and Gerardo P. Sicat, *Economic Policy and Philippine Development*, University of Philippines Press, Manila, 1972.
3. An ILO mission which surveyed the unemployment situation found open unemployment to be 6.1 per cent and also made a "guesstimate" of total unemployment (i.e. open unemployment plus an inadequate income measure of underemployment) in the vicinity of 25 per cent. International Labour Office : *Sharing in Development : A Programme of Employment, Equity and Growth for the Philippines* (Geneva 1974).
4. Based on a study on "Physical planning strategy for the Philippines" conducted by the Department of Public Works and Communications, the Institute of Planning of the University of Philippines and the

Presidential Advisory Council on Public Works and Community Development with the assistance of UNDP, three alternative regionalization proposals were worked out. Later, the Commission on Reorganization appointed an Inter-Agency Committee which chose the regions.

5. Normally the budgetary procedures are the most recalcitrant to reform. Therefore this regionalization of the budget must be seen as an innovation, for it paves the way for allocation questions on area basis rather than along sectoral lines, while ensuring flow of resources to the region. This structuring will facilitate horizontal programme integration and coordination and the subsequent monitoring of implementation. A similar innovation has been introduced for improving the District Planning procedures in India. See K.V. Sundaram : Case Study of Maharashtra in *Multi-level Planning for Integrated Rural Development in India*, Heritage, New Delhi, 1980.
6. Ed. B. Prantilla : Regional Development and Planning, the Philippine experience in Growth Pole Strategy and Regional Development Planning in Asia. *Proceedings of a Seminar* 4-13 November, 1975 United Nations Centre for Regional Development, Nagoya, Japan.
7. Rene Santiago writing in 1976 reported that six RDC's (Regions 5, 6, 7, 9, 10 and 11) had submitted their plans to the NEDA. See, Rene Santiago : Regional Planning and Implementation : The Philippine Experiment, Colloquium 76 (25 October—1 November, 1976), Methods of Planning for Comprehensive Regional Development by the UNCRD, Nagoya.
8. Rene Santiago : Regional Planning and Implementation: The Philippine Experiment. Ibid.
9. Ibid.
10. For a good account of the Integrated Rural Development strategy adopted in the Philippines, see : Carlos P. Ramos ; Institutional mechanisms for managing integrated rural development, the Philippines experience, Paper to the Round Table on Adaptation of Administration to Rural Development Organised by Government of India and ESCAP, 16-18 August, 1978, New Delhi (Mimeo).
11. See Carles P. Romos : Institutional mechanisms for managing integrated rural development, the Philippine experience, Ibid.
12. For an account of the Barrio Irrigators' Service Association, see the Chapter on 'The Philippines' in Local-level Planning for Integrated Rural Development—Report of an Expert Group Meeting Bangkok 6-10 November, 1978, United Nations, ESCAP.
13. These are archaic government procedures, civil service and auditing rules and budgeting procedures.
14. Ibid.

CHAPTER 7

THE NORTH-SOUTH PROBLEMS OF THAILAND

THAILAND (population 41.6 million in 1975 growing around 3 per cent during 1960-75)[1] is another example of a primary-product-based export economy in transition. Prior to the fifties, its development pattern was "basically autonomous, expansionist and satellitic"[2] characterized by the so-called outward looking policies within the framework of an inegalitarian distribution of property and incomes. Economic development planning has had a relatively late start, not until 1961,[3] and that too under pressure from outside agencies, particularly the World Bank and USAID[4]. The First and Second Five Year Plans covered the periods 1961-66 and 1967-71 respectively. They were primarily nationally oriented, aiming at the rationalization of public resources allocations and the promotion of vigorous private investments. Under these plans, the public sector investment objectives concentrated on the construction and rehabilitation of basic infrastructural facilities such as highways, electricity and irrigation systems, schools and hospitals. The resultant economic growth during the 1960s was quite remarkable, partly owing to the successful achievements of the first two plans and partly due to favourable international economic conditions as well as internal political stability. During the period, the gross domestic product at constant prices grew at an average rate of 8.1 per cent annually and per capita income almost doubled.[5] Despite this remarkable national economic growth and upward change in the level of per capita income, regional gaps in income and production widened. Although the Second Plan contained a general formulation for the "promotion and maintenance of social justice and the provision of relief to people in isolated areas",[6] it was since the Third Five Year Plan (1972-76) that concern for the planning for sub-national regions became explicit and articulate. The current Fourth Plan (1977-81) contains clear-cut development guidelines aiming to reduce the existing socio-economic disparities and to accelerate improvements in rural incomes and employment opportunities.

In its regional structure of underdevelopment, Thailand is a typical instance of the contradiction reflected in the metropolis-satellite position. In all respects, Bangkok illustrates most aptly Mark Jefferson's primate city[7] with all the political, economic, social and intellectual life of Thailand concentrated in it. The growth and dominance of Bangkok has generated not only inter-regional disparities in the level of economic welfare but also undesirable and unbalanced distribution of population. Greater Bangkok has about 10 per cent of the total population of Thailand, that is, about 4 million; it has about 57 per cent of the total urban population of the country. Its population is more than 20 times larger than that of its next largest urban centre that is Chiengma. During the 25 years of post-war period from 1947 to 1971, the metropolitan population increased by 300 per cent* from 781,660 people to 3,075,300 people and the total urbanized area grew by 244 per cent from 53 Km^2 to 182 Km^2.

Thus the Bangkok-Thonburi area represents a metropolitan agglomeration and the rest of the country a rural hinterland of primary produce. This rural hinterland may be divided into four regions—the north, north-east, south and central plains—each of which has a substantial economic identity. The Bangkok region falls in the central plains. The extent of polarised development and inter-regional disparities are reflected in the GDP figures by regions. The following Table 7.1 is meaningful.[8]

* Table provided in Phisit Pakkasem : Industrialisation strategies and the Growth Pole approach to South Thailand Regional Planning : Paper included in "Growth Pole Strategy and Regional Development Planning in Asia *Proceedings of a Seminar at Nagoya*, Japan 4-13 November, 1975, United Nations Centre for Regional Development, Nagoya, Japan.

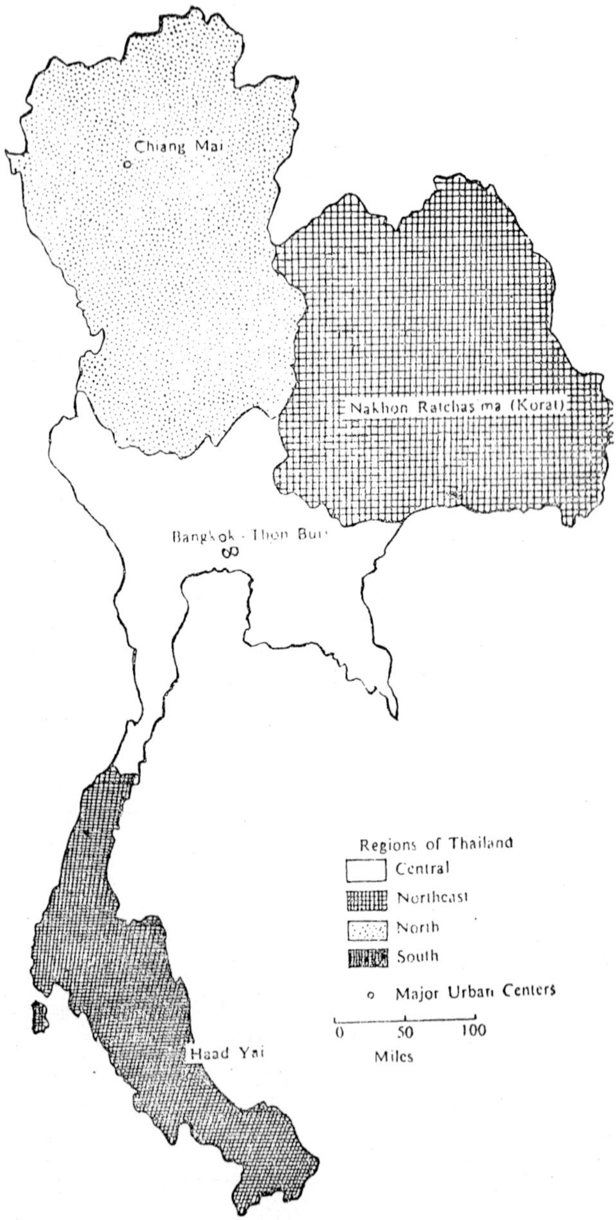
Chiang Mai
Nakhon Ratchasima (Korat)
Bangkok - Thon Buri
Haad Yai
Regions of Thailand
Central
Northeast
North
South
Major Urban Centers
0
50
100
Miles

FIG. 4

TABLE 7.1 : **Summary of GDP by regions, 1973**

	Whole Kingdom	North-East	North	Central	South
Gross Domestic Product (Million Baht)	187,681	30,206	27,184	107.912	23,319
GDP per capita (Baht)	45 40	2 143	3 024	8 230	4 343

Source : National Accounts Division, NESDB

The above table indicates clearly that both economic activities and incomes are heavily concentrated in the central region, especially in the Bangkok area. The north and the north-east are the most depressed, with the south maintaining a level of per capita income somewhat equal to the national average.

Economic trends and national policy-level thinking undoubtedly point to some concern for regional planning and the application of some concerted efforts towards regional development. Sub-national approaches in development planning and regional development strategies are still evolving. In this dynamic context and particularly in a situation where development performance has yet to come to terms with regional potentials and problems,[9] and with a relatively short span of experimentation with regional planning, the question as to whether regional disparities are increasing or decreasing is not perhaps quite relevant. All that is significant to note is that while the growth rate for the central region is being sought to be decelerated, those of the others are being accelerated.

The following Table 7.2 is illustrative of this trend.[10]

TABLE 7.2 : **Growth rates in GRP in per cent per annum at constant (1962) prices**

Period	South	Central	North	North-East	Thailand
1960-67	5.8	8.7	6 9	6.6	7.7
1967-70	7.8	7.7	6.8	7.5	7.6

While the contrasts in levels of development achieved in Thailand's different geographic regions constitute one dimension of uneven development, the disparities in income distribution and proverty levels among the people reflect yet another dimension of inequality. The number of people in poverty has been estimated at 13 millions in 1975, which is 32 per cent of the population. The share of the lowest 40 per cent of population in the GNP has been estimated at 11.5 per cent (1975).[11] Both these figures are comparable to the situation obtaining in the Philippines for the same period.

In order to effectively capture the essence of the problem of the spatial and income distribution inequalities characterizing Thailand's economy today, it is necessary for us to note that the present structure of underdevelopment in the country is somewhat different from the pattern of colonialism common to most other developing countries in South East Asia. Thailand is perhaps unique because it has remained independent throughout the last two hundred years with no period of its history under formal colonial rule. Therefore, whatever internal differentiation in its development has taken place is an inheritance from the policies pursued by its own governments. The results of such an inheritance, however, have been no different from those characterizing countries under long colonial rule; for it partakes all the characteristics of colonialism in the economic front, its economy has always been outward-looking, responsive to the foreign demand for agricultural and mining products and its interior shows all the features of spatial penetration and not spatial development.

Thailand's development pattern in the century preceding World War II has been characterized as stemming from the laissez faire policy of the Thai government. Muscat has concluded that it was "basically autonomous, expansionist and satellitic", having been largely "a response to external demand, with a much smaller stimulus from the domestic market in Bangkok". This led to vigorous expansion along lines of comparative advantage that is, in the production and international trade of the indigenous primary products—tin, teak, rubber and rice. The comparative advantage principle is a strategy that we advocate in this book for the development of backward areas. But this strategy has to be applied under certain controlled conditions. Basically, it should be supported by a policy of plugging certain kind of leakages to the maximum. Only then can the advantages in following such a policy accrue to the backward areas. In the case of Thailand, the comparative advantage doctrine was exploited *primarily* as an engine for growth transmission connecting the country with the centres of industrial (and population) growth, and only secondarily as a means for improving the allocation of existing resources. The point that we would like to emphasize here is that the adoption of the comparative advantage doctrine was not a wrong policy in itself: it was technically inadequate in its application so far as the development of backward areas in the country was concerned.

During the decade 1960-70, the Thai economy received an unexpected boom. Among the reasons that contributed to this boom were an increasing external demand for primary products, a bumper crop of paddy, increase in public and private investment, a large increase in unrequited transfers (representing mainly the earnings from illicit business activities involving South Vietnam) and the inflow of foreign capital (mainly from the rapid build-up in US military expenditures in Thailand). This boom naturally brought with it considerable foreign receipts, which if they had been more prudently channelled, would have taken the Thai economy to greater heights along the path of more even development. But these foreign receipts were frittered away more in conspicuous spending in Bangkok to maintain the primate city's pseudo-affluence than for initiating a process of balanced development in the country. When the boom was on, the increased demands for primary products, for example paddy, were met through frontier expansion rather than through increased productivity.[12] With such a pattern of development, the hinterland of primary produce in Thailand stagnated with low productivity and low incomes and consequently with poverty and underdevelopment. Now there are no outside economic forces capable of breaking the "low productivity-low income-low investment circle" in which these areas are bound. This, in brief, is to put the context of the geography of underdevelopment in Thailand in its proper perspective. With the cessation of the Vietnam war and the drop in foreign exchange inflows, an irreversible change in external political events has occurred which must be reckoned as a reality while interpreting the outlook for regional development in Thailand. Undoubtedly in the economic history of the nation, an opportunity for the spatial restructuring of its economy has been missed. Therefore, despite its past accretions of foreign exchange and impressive growth rates, Thailand starts on a low base in its sub-national (regional) development front, with no great initial advantages to bank upon.

It would be unfair to characterise the attitude of the Thai government as a total lack of concern for the development of the less developed regions. As indicated before, active governmental intervention for development planning itself has had a relatively a late start in Thailand and in this context the concern for sub-national development was a further later development. Since the Third Five Year Plan, increasing attention is being directed towards the development of depressed regions in the country. The Fourth Five Year Plan has stated the reduction of regional disparities as one of the four national development objectives. According to Pakkasem, regional development strategies occupy high priority among its *ten* main national strategies of the Fourth Plan.[13] We must add here that the perception of the disparities problem was considerably accelerated by the recurring economic and political unrest, including insurgent activity that occurred in some parts of the country since the early 1950s, particularly in the light of events in neighbouring Vietnam, Cambodia and Laos.[14] Governmental interest came to be focussed on the problem since then, and in fact, some transfer of resources to some of the lagging regions did take place during the first two national plans.[15] However, these were inadequate and did not contribute much towards the amelioration of the problem. The experience in itself is however worth studying for the insights it affords into the problem of reducing interregional differences in an underdeveloped country. During the 1970s the Thai government commissioned some studies to identify structural problems and solutions for various regions and sub-regions in the country. They include a regional planning study for the North-east undertaken with the technical assistance of a USAID team and another for the South undertaken with the assistance of the U.K. Government through Hunting Technical Services Limited. There is an excellent review of these studies by Pakkasem.[16] The experience of the north-east has also been studied by Dixon.[17] A recent book by Wolf Donner *The Five Faces of Thailand*[18] is a good geographical study and sets out the problems and potentials of the various regions. Using these source materials, we may build some further insights into the regional problem of Thailand.

The north-east is the most backward region of the country with low per capita income (per capita income less than half the national average as in 1973), limited transport facilities, poor soils, irregular distribution of water and severe weather. Its underdevelopment often attributed to the environmental (geographical) factors has been questioned by analysts, who have rightly argued that historical, economic and political factors have equally contributed to it.[19] During the post-war period, repeated incidents of unrest in the north-east linked with the events in neighbouring Indo-China caused considerable concern. This led the government to take up some adhoc measures by implementing some piecemeal development schemes aimed at transport, health, education, water supply and irrigation. These measures, however, could not contain the unrest; the matter became quite serious, as terrorism and violence started in the mid-1960s. This led the Government to adopt from 1961, a more coordinated development policy with substantial assistance from USA as part of the first national development plan. Following these developments, a North-East Economic Development Committee under the Chairmanship of the Deputy Prime Minister was created to prepare a comprehensive plan for the north-east. Under the aegis of this Committee, a North-East Economic Development Project was formulated with the assistance of a consultant team provided by USAID. This plan turned out to be a summation of the public investment projects proposed by various central operating agencies to the National Planning Authority, and not strictly those originating from an analysis of the region's needs and growth potentials. In effect, it was only a regionalization of development expenditures along the traditional lines of sectoral allocation without consideration of spatial efficiency. Thus the results of these studies were "very weak and not up to the standard that the Thai government or the Regional Development Committee expected".* Between

* The regionalization of developmen texpenditures in a mechanical way, retaining the top-down process of planning negates the whole concept of (*Contd.*)

1961 and 1971, the policy for the north-east stressed the development of infrastructure, particularly transport and irrigation and to a lesser extent health, welfare, education and agriculture. This programme was sought to be combined with rural development schemes and propaganda stressing the evils of communism. The guiding philosophy appeared to be basically political namely, "If stomachs are full, people do not turn to communism." All these, however, failed to generate a reversal of the regional growth trends[20]. According to Dixon's analysis :

> Heavy investment in irrigation has failed to produce any worthwhile return; roads were geared to security rather than development and a whole series of schemes were primarily designed to increase the government's popularity in sensitive areas. Much of the health, education and welfare expenditure can be considered as lying in this category, and the extension of rural electrification and village road programmes has largely been to areas where security problems are rapidly worsening. Such schemes generate only small increases in regional incomes. The very large political element in the policy directed towards the north-east has resulted in heavy expenditure on propaganda and security, and has severely reduced the impact of the development policy and on the growth of the region's economy.[21]

The case study of the north-east shows that in spite of the Thai government treating the north-east as a unique problem area and directing efforts towards solving the multifarious problems of this region, the results have not yielded any

regional planning, as it does not proceed logically from an analysis of needs and potentials and seek to match these with resources allocation from the Centre. It also fails to influence substantially the criteria for inter-regional allocation of resources. However, it cannot be denied that in developing countries facing several limitations for initiating a systematic process of regional planning, even a mechanical disaggregation process of this kind does assist to provide an understanding of the region-wise quantification of the national development effort. This, in turn, may lead gradually to building up pressures among the regions for more rational allocation procedures.

remarkable success in terms of closing the inter-regional gap. However, if it is agreed that, regional development is a long term process, one may argue that the time taken by the Thai government to initiate these developments is really to be interpreted as the lead time necessary to create a framework for development designed to control and order a more rapid process of social and economic change. If this interpretation is correct, the next stages in planning will be most crucial. It remains to be seen how the Thai government is going to assign priorities, practise some trade-offs, and in short, manage and carry out the several important adjustments that are called for in its altered context of political and economic circumstances* to reach the goal of balanced development among the regions.

The regional studies pursued in the context of the southern region of Thailand also deserve our attention, as they point to certain strategies that are relevant in tackling the problems of lagging regions. The southern region accounting for about 12 per cent of national value added and a similar share of the nation's population has per capita income levels almost roughly equal to the national average. While rice and rubber cultivation are the major primary activities, it has also other endowments to support a diversified primary base which have been identified in the following directions : commercial fishing, forestry, mining, fruit and other high return crops and activities. The level of industrialization is low and the fluctuations in international commodity prices (particularly tin and rubber have contributed) to its relative backwardness. In undertaking a study of this region, the Hunting study team has (a) looked at the regional economy in a national context and attempted to identify sectoral trends and their inter-relationships within the economy; (b) taken note of the physical constraints and opportunities within the region; (c) formulated appropriate sectoral strategies and identified projects to promote these

* The erosion of the external finances, producing an irreversible change in external political events has altered the political and economic situation in Thailand. One view is that when the Thai government could not make the desirable adjustments in better times when foreign exchange was plentiful, how can they expect to carry out these under more painful and less manageable conditions?

strategies; (d) made necessary futuristic studies to assess the major requirements and demands, and (e) evaluated the identified projects in both qualitative and quantitative terms to pre-feasibility stage.[22] While this methodology is interesting, what is relevant in our present context is the kind of strategies proposed by the team to develop the lagging region. The report points out that direct assistance to depressed areas may not always be the best way of helping them. As examples of indirect policies that assist the relatively poor areas, policies designed to accelerate a process of urbanization and those of developing new agricultural areas have been mentioned. On a balancing of various considerations, the team has recommended a policy of concentrated dispersal of investments in the region. In the words of the team's report :

> We favour the concentration of effort in the most exploitable resources of each province or area. By concentrating particular activities, economies of scale and of comparative advantage can be obtained. By dispersing the various activities, it should be possible to ensure the advantages of a fairer distribution of benefits. Our aim has been therefore to develop a balanced package of proposals and programmes for the whole region, biased where desirable in favour of the poorer areas, but in all cases aiming to fulfil as far as possible each area's potential......Such an approach cannot be expected to lead to dramatic changes in the relative position of the poorer areas, but in the longer term we believe that much can be done to control the extent of income inequalities and that this is best achieved in the context of balanced regional development.[23]

The third lagging region of Thailand—the north, has about one-third of the area of the country and just over one-fifth of its population. Its average per capita income is about one-third less than the national average. A large portion of its area is mountain and forest, particularly of teak trees which are an important resource of this region. It has also some 500,000 tribesmen. The population of the region is growing at a high rate approximately 3.1 per cent. It is a heterogeneous region characterized by physical, economic and social diversity. With about 22 per cent of the country's population, the region's contribution is around 15-16 per cent of the GDP of Thailand. In 1971-72, under a joint Thai/UN sponsorship, the Institute of Planning and Development (IPD) Tel Aviv was engaged to study the development problems of the North.[24] This study recommended the development of a growth area, comprising Chiang Mai, Lamphun and Lampang. A three-stage strategy in the development of this growth area could be by (a) diminishing the growth area leakages through reduced extra-regional transfer payments; (b) increasing the value added in the growth area from output by encouraging forward production linkages; and (c) encouraging the return of leakages by promoting increased government operating expenditures in the growth area and a higher rate of government and private investment. Other recommendations of this study group include: (1) regional planning should begin at the base, that is, the Changwat (province) and be moved away from centralised decision-making, (2) agricultural production should be intensified and diversified, and (3) incentives should be given to promote industrialization outside the area of Greater Bangkok.

Following the above baseline studies for the north-east, north and south, more attention is being directed towards the development of depressed regions in the country. The policy guidelines set out in this behalf in Thailand's Fourth National Plan are being translated into a set of regional and spatial strategies, policies and action programmes for sub-national development purposes. Briefly there are three major development strategies that are influencing the development of these peripheral regions today. These are :

Strategy for the intensification and diversification of the economy for improving regional industrial mix and export base

The phase of expansionist growth has almost come to an end for Thailand and a phase of intrinsic growth must now take its place. Basically, this would consist of seizing the comparative advantage principle in development and organising and strengthening inter-regional economic relationships via regional specialisation, diversification and trade. This is not an easy task and would call for several organisational and administrative adaptations and innovations, for which there are always certain definite limitations in every developing country.

Strategy for decentralized urbanization

This implies reducing the primary effect and industrial pull of Bangkok and promoting decentralisation of industrial activities and regional urban or growth centres for all the lagging regions—the north-east, the north and the south. This would imply promoting urban infrastructural development programmes, setting up industrial estates, developing inter-urban accessibility network, ports and overland road networks. The specific regional studies have identified the growth centres to be promoted.

Strategy for decentralised infrastructural and social services to the rural areas

In keeping with the thinking to promote the development of the lagging regions, a new economic infrastructural policy has been evolved. This policy favours the smaller and rural-oriented infrastructural development programmes in public utilities, transport and electrification as well as promotion of social services, particularly health, education, and welfare programmes. These are intended to narrow the service gaps between the rural and urban areas, and between the rich and poor communities.

Parallel to the initiation of regional development effort, some thinking on local development has also started, although this is a relatively recent experience. The Changwat (Provinces), Amphoe (Districts) and Tambon (Sub-districts) are the units at the subnational level to which the planning process has to be ultimately decentralized. Such a development in multi-level planning would take long time to fructify. The Changwat (Provincial) planning offices have been set up for all major provinces in the country. This office is attached to the office of the Provincial Governor and is guided by a Provincial Planning Committee (PPC) under the Chairmanship of the Provincial Governor with some 20-25 departmental representatives of field level and representatives of local self-governing bodies. The PPC is vested with the development planning functions and has responsibility to coordinate and integrate both the top-down inputs and the bottom-up basic felt needs of villages, communities and districts. At the national level, the Central Committee for Provincial Development (CCDP) (which has also links with the National Economic and Social Development Board) is vested principally with the coordination and integration of provincial planning and development activities along the lines of the regional and national objectives and policies. These developments are relatively recent ones. Undoubtedly, Thailand has to go a long way to make the bottom-up process of planning a reality. At present, most of the local level planning developments such as the Tambon Schemes, the Provincial Planning Schemes, Projects of the Department of Community Development such as Saraphi and certain special projects of the Land Settlement Division are all government initiated and executed. To make the local level planning process effective, apart from clear-cut policy directives, appropriate institutional mechanisms as well as special financial supports from the Government are needed. An important development in this context initiated since 1979 is an outright grant called Changwat subsidy fund—made available to Changwat Administrative Council for financing local development projects in the Changwat plan.* This fund is a special development grant on top of the normal national subsidies to the local government. This is intended to strengthen the role of local governments in their development planning and implementation.

The rural development strategy in Thailand follows essentially a top-down approach and is concentrating mainly on the provision of necessary inputs and resources to the local areas through extension strategies that is, from the urban centres which are the seats of knowledge, of technology and of capital. Development programmes based on such strategies have failed again and again because they had ignored a number of very important factors such as (a) the social, political and cultural aspects of the local community as well as the macro organisation of the society, (b) the differences

* In the fiscal year 1979, the Government earmarked one per cent of the national budget for provincial plan implementation process.

between each community, area and even region as regards needs and absorption capacity, and (c) streamlining the organizational aspects of the delivery system.

In summing up, it may be noted that a number of regional and local development policies and strategies have been initiated in Thailand. These are relatively recent phenomena. By their very nature, they are long term processes. In the context of the several constraints operating today in Thailand, their impact can be felt only in a gradual fashion with cumulative effects spread over a number of years. Careful and sustained planning informed by a proper appreciation of priorities would determine the pace and pattern of development in the future years. Political stability and conviction in the scientific method are essential if the policies outlined by the government in the Fourth National Plan are to be purposefully carried so as to bring new opportunities for the poorer regions of the population and to narrow down the interregional gap.

NOTES AND REFERENCES

1. Growth and Poverty in Developing Countries—World Bank Staff working paper No. 309 (Revised) May 1979, Tables 1 and 2.
2. Robert J. Muscat; Development strategy in Thailand, a study of Economic Growth—Praeger special studies in International Economics of Development, New York, 1966.
3. According to Muscat, "the felt need for development has appeared at the top of Thai Society, not as an expression of discontent on the part of sections of the elite dissatisfied with reward of the politico-economic system, but as a response to growing external political threats in South-East Asian Peninsula."
4. The World Bank Mission report observed : "Public development effort has been unbalanced, uncoordinated and excessively diffused. There is consequently a very real need in Thailand for a central agency to make continuing study of the nation's economy and to draw up plans for its development". International Bank for Reconstruction and Development; A Public Development Programme for Thailand" (Baltimore : The Johns Hopkins Press, 1959).
5. National Economic Development Board : National Income of Thailand; Bangkok Government House Printing Office, 1971.
6. Government of Thailand, The Second National Economic and Social Development Plan 1967-71, Bangkok, 1968.
7. Mark Jefferson, The law of the primate city. *Geographical Review*, 29, 1939.

8. According to Phisit Pakkasem, "the population of Metropolitan Bangkok had increased by 6.2 per cent during 1960-70 and that its growth exceeded the natural rate of increase by 3.3 per cent per year." See Phisit Pakkasem. Thailand country Report to UNCRD Seminar on 'Urban Development Strategies' at Nagoya, Japan, 28 Oct.-8 Nov. 1974.
9. Although the era for expansionist growth in Thailand is more or less coming to a close, there are still some areas in the nature of undeveloped or underdeveloped lands that present a great potential for intensification and diversification of the economy.
10. These figures are quoted from a paper presented by Arthorn Suphapodok to a seminar on "Role of cities in attaining desirable population distribution in the context of rapid urbanisation" organised by the United Nations Centre for Regional Development, Nagoya (Japan) Oct. 28-Nov. 8, 1974 entitled "South Thailand, A study of settlement and migration patterns" (Mimeo). The periods chosen here for comparison are considered to be distinct periods in the Economic History of South Thailand.
11. *Source* . Tables 1 and 2 in World Bank Working Paper No. 309 (Revised) "Growth and Poverty in Developing Countries", May 1979 (Mimeo).
12. According to recent assessment, this period of expansionist growth has more or less come to a close. Now Thailand has to build up its economy on the principle of intrinsic growth. The road to such intrinsic growth is not easy, as it requires continuous improvements in technology and organisation, improvements more difficult to bring about than mere extension of already known and accepted techniques and organization to an expanding frontier.
13. Phisit Pakkasem, Country Report on Thailand, in the Colloquium on Methods of Planning for Comprehensive Regional Development. Oct. 25-Nov. 1, 1976, United Nations Centre for Regional Development, Nagoya (Japan).
14. J. Nakahara and R. A. Witton, *Development and Conflict in Thailand*, Cornell University Press, Ithaca N.Y. 1971. Also D.E. Nuechterlein, *Thailand and the Struggle for South-East Asia*, Cornell University Press, 1965.
15. According to Pakkasem, during this period, the North-east absorbed 28 per cent of the total budget, a percentage only slightly less than its share of the population. See P. Pakkasem, *Thailand's North-East Development Planning : A Case Study in Regional Planning*, National Economic and Social Development Board, Bangkok, 1973.
16. P. Pakkasem : Country Report—Thailand, Ibid.
17. C.J. Dixon, Development, regional disparity and planning, The experience of north-east Thailand, *The Journal of South-east Asian Studies*, Vol. VIII, No. 2, Sept. 1977.
18. Wolf Donner, *The Five Faces of Thailand*, Institute of Asian Affairs, Hamburg, St. Martin's Press, New York, 1980.

19. See C.J. Dixon, Ibid, p. 222. Also Wolf Donner, *The Five Faces of Thailand*, Ibid, pp. 547-548.
20. P. Pakkasem, Country Report : Thailand, Ibid.
21. C.J. Dixon, Development, Regional Disparity and Planning : The experience of north-east Thailand. Ibid.
22. The report of the team is in three volumes: (1) Resources and strategies, (2) Physical planning, and (3) Implementation. See, South Thailand Regional Planning Study made by Hunting Technical Services Limited under the sponsorship of the Government of the Kingdom of Thailand, National Economic and Social Development Board and the Ministry of Overseas Development, London.
23. South Thailand Regional Planning Study Vol. I, p. 11, Ibid.
24. UNDP : Planning for the Northern Region : Project findings and recommendations, DP/UN/THA-69-533/1, New York, 1974.

CHAPTER 8

UNDERDEVELOPMENT IN INDONESIA

THE DISPARITIES problem of Indonesia is sometimes described as a plenum and vacuum pattern of contrast between Java and the Outer Islands. Among the Islands of Indonesia, Java is comparatively resource poor, where however population and economic activities are heavily concentrated. The Outer Islands have rich natural resource endowments, but the economic activities of these Islands are still confined to a raw material extraction and export stage with very low population density, so that in many ways they serve as the hinterland of Java. Thus the relations between Java and the Outer Islands seem to flow from a core-periphery structure.[1] The following observation of Geertz seems quite significant :

> If ever there was a tail which wagged a dog,
> Java is the tail and Indonesia, the dog.[2]

Population density differences between Java-Bali and other Islands are quite striking and this maldistribution of population seems to be the major single factor underlying the regional problem of Indonesia. The population of Indonesia has increased remarkably during the present century from about 48.5 million in 1920 to 97 million according to the census in 1961 and to 119.5 million in 1971. A World Bank estimate for 1975 puts it at 130 million. According to 1971 census, 76 out of the 119 million or 63.8 per cent (nearly 2/3rd) of the population was in Java-Madura Islands that is, concentrated in about 6.6 per cent of the land area. In density terms, Indonesia as a whole has 59 persons per square kilometre. Java and Madura has about 1000 inhabitants per square kilometre; on the other hand, the Outer Islands have a density of about 20 inhabitants per square kilometre.

The population pattern is also reflected in land utilization. While the proportion of cultivated land in Java is as high as 70 per cent, it is only 4 per cent in the Outer Islands. The intensive cultivation in Java is predominantly of the wet rice agriculture pattern, while the predominant pattern of farming in the Outer Islands is swidden agriculture, that is, shifting cultivation or slash and burn farming.

The country is predominantly rural with only 17.4 per cent (about 22 million in 1971) living in the urban areas.

Indonesia is one of the poorest countries in the world, ranking 14th from the bottom, according to a report of the world Bank. The per capita GNP is only US $ 90 annually (1975). Statistical information on the distribution of income and wealth in Indonesia, as is the case for most developing countries, is scarce. The general impression of a high concentration of both income and wealth in the hands of a small minority of the population, most of whom are upper class urban dwellers is confirmed by available evidence. According to the estimates of size distribution of income compiled by Shail Jain[3], the percentage of income received by the top 5 per cent of the households in Indonesia in 1971 was 34 per cent which is the highest among the group of five countries that constitute our study area. According to a 1975 estimate reported in a World Bank publication[4], the share of income of the lowest 40 per cent of population is 16.1 per cent. Another measure of overall equality is the number of people below the poverty line. The percentage of population in poverty in Indonesia in 1975, (using Kravis adjustment factors) has been computed as 59 per cent[5], which is again the highest for Indonesia among the group of five countries. A study by Leiserson *et al*[6] gives the breakdown of income distribution by rural and urban areas. According to this study (which takes per capita consumption expenditure of Rs. 3000 per month as the threshold), the incidence of poverty in the rural areas was 50.2 per cent overall. Rural Java accounts for the highest incidence of rural poverty with Central and East Java having 67.8 per cent and 66.3 per cent respectively. The range among the different provinces is quite wide. Kalimantan has the lowest incidence, the proportion being 17.3 per cent. For urban areas, the proportion of those in poverty was 18.7 per cent in Jakarta and 33.4 per cent in Central Java. The findings of more than one survey have nevertheless indicated that particularly in the urban areas of Java, there was a considerable diminution in the numbers of those in poverty between 1970 and 1976. This reduction has occurred during a period of comparatively rapid economic expansion. However, the Lieserson study also points to evidence indicating that overall inequalities in income in the country have continued to grow, the urban-rural disparity has widened, and within the urban area the benefits of rising productivity would seem to have accrued more to profits and property incomes than to real wages.[7]

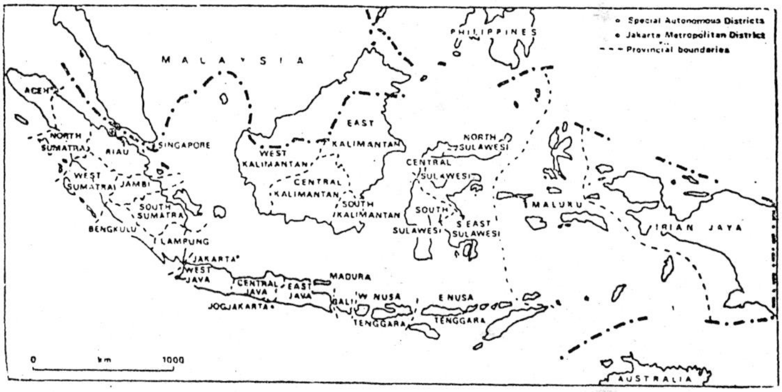

FIG. 5

While unevenness in development among the different provinces exists to a marked degree, there is very little comparative quantitative information available on levels of development. A study group on regional income set up under the chairmanship of Hendra Esmara[8] in 1970 has, however, provided some estimates for the years 1968-72.

In Indonesia, regional inequalities are quite striking in regard to natural resource endowments, in the stock of man-made capital per head, in the quality of labour and in the efficiency of resource use and organisation. In natural resource endowments, regional inequalities are conspicuous by the concentration of oil and timber resources in a few regions, for example, Riau, East and Central Kalimantan. In fact, when the Regional Gross Domestic Product (RGDP) per head is calculated, taking the income generated from these resources, it tends to give a highly distorted picture. The regional income group therefore excluded oil income from the GDP calculations. The following Table 8.1 brings out the inter-provincial income disparities for 1972 taking the per capita GDP (excluding oil) as a measure :

TABLE 8.1 : **Provinces classified according to GDP**

Per Capita GDP		Provinces
(Excl. Oil) ('000s)		
15—20	—	West Nusatenggara, (18) East Nusatenggara (20)
20—25	—	West Kalimantan, Yogyakarta South Sulawesi, Central Sulawesi North Sulawesi.
25—30		Central Java West Java South East Sulawesi West Sumatra Bali
30—40		Lampung Bang Kulu South Kalimantan East Java Irian Java Aceh
40—50		Maluku Jambi North Sumatra South Sumatra
50—75		DKI Jakarta (55) Central Kalimantan (63)
75—100		Riau (88)
100—200		
200—250		East Kalimantan (222)

Source: Hendra Esmara, Regional Income Disparities, *Bull. Indonesian Econ. Studies* Vol. XI, No. 1, March 1975.

Thus it may be seen that in terms of per capita GDP, East Kalimantan leads, followed by Riau and Central Kalimantan and the least per capita GDP is to be found in West Nusatenggara. The regional gap in per capita GDP among the provinces is 12 times, which shows that regional disparities are truly enormous. The high per capita GDP of East and Central Kalimantan reveals the influence of the timber industry. In the case of Riau, the high per capita GDP, even after the exclusion of oil income is because of the impact of the oil industry on other sectors. The high degree of regional inequality in Indonesia is accounted for by the three provinces, East and Central Kalimantan and Riau. The coefficient of variation for the country as a whole is 0.522. When the three provinces are excluded from the calculation, the value falls to only 0.262.[9] There are also wide differences among provinces in urban consumer prices. Such wide inter-regional difference in price levels reflect transport costs and other obstacles to the flow of inter-regional trade. They are a measure of the extent to which Indonesia still consists of interrelated regional economies rather than an integrated national economy.[10]

Developments are mostly concentrated in the Islands of Java, Sumatra and Kalimantan. The following Table 8.2 summarising the behaviour of the coefficients of regional inequality by sectors serves to underline the extent of spatial concentration:

TABLE 8.2 : **Coefficients of regional inequality by sectors**

Sector	*Coefficient*	*Remarks*
Agriculture (incl. forestry)	0.842	0.288 when Jakarta Central and East Kalimantan excluded.
Industry	0.422	
Transport	1.409	0.369 when Riau, South Sumatra, Jakarta and East Kalimantan excluded. 0.352 when Riau, Central and East Kalimantan excluded.

The distribution of manufacturing took whether measured in terms of the number of establishments, employment or value added shows excessive concentration in Java. The following Table 8.3 is self-explanatory :

TABLE 8.3 : **Regional distribution of manufacturing by major regions**

Indicator of Development	Percentage distribution in					
	Java	Sumatra	Kalimantan	Sulawesi	Other Island	Total
Total Establishments	74.8	14.5	3.0	5.4	2.3	100
Total Workers	85.1	10.6	1.4	1.5	1.4	100
Total value added	80.3	16.6	1.3	1.0	0.8	100

Source : Statistik Industri 1971 quoted in Hendra Esmara Regional Income Disparities (ibid).

Although a 5-year period (1968-72) is too short for any measurement of trends, certain conclusions seem obvious :

(*i*) There has been marked tendency for regional income disparity to increase, such disparity arising primarily from resource extraction (oil and timber production) in the three provinces mentioned earlier.

(*ii*) The resource rich regions are on the average growing faster than the comparatively resource poor regions. Thus the growth rate of GDP at 1969 constant prices during 1968-72 was an aggregate of 14.4 per cent for the resource rich provinces of E. Kalimantan, Maluku, Irian, Java, Riau, S.E. Sulawesi and Aceh, while it was an aggregate of only 6 per cent for the provinces in Java and Bali-islands.

(*iii*) The diverging trends among the regions become obvious when they are cross-classified according to their incomes (per capita RGDP) and growth rates. Table 8.4 illustrates this trend.

From Table 8.4, it may be seen that most of the population (51%) is in the relatively low income and slow growth category. About 15 per cent of the population is in the high income—high growth category. These two groups account for the increase in regional disparities. Only a very small part of the population (5%) is in regions with low

TABLE 8.4 : **Distribution of regions by per capita RGDP and annual growth rates**

	High Incomes	Low Incomes
High Growth	7 Regions: East Kalimantan, Riau, Jakarta, Irian, Java, North Sumatra, Aceh Maluku Share of Population: 15% Share of GDP : 25%	3 Regions: Bali, West Sumatra, Southeast Sulawesi Share of Population 5% Share of GDP 4%
	6 Regions: Central Kalimantan, Jambi, South Sumatra, South Kalimantan, Bengkulu, East Java	10 Regions: Lampung, West Kalimantan, Central Sulawesi, North Sulawesi, South Sulawesi, West Java, Central Java, Yogyakarta, East and West Nusatenggara
Low Growth	Share of Population : 29% Share of GDP : 31%	Share of Population 51% Share of GDP 40%

income and high growth rates. Nearly 1/3rd of the population (29%) is in regions with high incomes but slow growth rates. If these trends are allowed to persist, it is clear that over time, disparities in income will tend to be very large.

The foregoing analysis of regional disparities in Indonesia clearly reveals that developments in the country have had their varied impact on different regions, in some cases reducing the initial disparities, in others accentuating them. It is also apparent that developments, as have taken place in the other Islands (other than Java) are largely confined to modern extractive industries of an enclave character. They have served, by and large, to merely exacerbate the dualism between the modern sector and the traditional sectors. Much of the income produced by the enterprises in the modern sector accrues to foreign companies, state enterprises and to Indonesian owners residing outside the producing provinces. Also, a large share of the income accrues to the Central Government in taxes and royalties, which is then transferred through public finance measures to support the economic development and the current living standards in ever more crowded Java. Thus the resources of the peripheral regions are being used to maintain Java's static development, a glaring example of internal colonialism. This has effectively throttled the regional development of those areas, wherefrom the income originates. Obviously conscious and effective policies are called for, if regional disparities are to be reduced and balanced development of the different regions is to be ensured.

Government policy measures

In Indonesia, conscious regional policies to correct the lop-sided development of the regions appear to have been pursued somewhat seriously only since REPELITA II that is the Second Five Year Plan (1974-79). This plan elaborated a series of regional development objectives.

> Since the national development plan is based on the sectoral development policy, every effort will be made to balance the sectoral with the regional development. Any sectoral development activities as far as possible will be linked towards an integrated regional development. However, particular attention has been given also to the *minus regions* (referring to the rice deficit regions), border regions and the backward regions.[11]

The regional development problems of Indonesia have been sought to be tackled by the government in two major directions :

(1) Through population redistribution and settlement policies which seek solution to a problem of ecosystem, characterized by uneven distribution of population and settlements. This is attempted through Transmigration schemes which may be seen essentially as a rural development programme in which the aim is to utilize the agricultural potential of the less populous areas by providing the manpower needed for agricultural expansion.

(2) Through a rural development policy in which sectoral, regional as well as Presidential grants-in-aid projects are directed towards the development of rural areas. It is a bottom-up strategy which is sought to be sustained through a mix of self-help and Government efforts.

The transmigration programme which originated as a measure to tackle the problem of poulation imbalances in the country is a unique Indonesian experiment in social engineering. It started as early as at the close of 19th century as a programme of colonization by the Dutch to arrange the movement of people from densely settled parts of Java to the largely empty Outer Islands. The rate of movement during this period was substantially below expectations (never more than 60,000 persons in a year). The total number migrated between the start of the programme and the Japanese occupation (1942-49) was under 200,000 and there was a huge gap between the number leaving Java each year and the annual increase in the Island's population.

In 1950, the new independent government continued the transmigration programme under a special department. Now the programme was seen not merely as a means of easing the plight of Java's landless peasants, but also as a method of furnishing labour to the Outer Islands. Although more colonists were moved between 1951 and 1960 than the Dutch had resettled in 40 years, the numbers involved, however, were still totally insufficient to have any impact on Java's worsening problems. The Eight Year Development Plan (1961-68) included an immensely ambitious proposal to move 390,000 families (about 1.5 to 2 million people) by 1968, but the performance fell far below the targets.

Since the beginning of the First Five Year Plan (1969-73) transmigration has been placed by the government in the wider context of regional development. The Second Five Year Development Plan (1974-79) proposed to move over 450,000 families from Java to Bali.

Critics have pointed out that the transmigration programme has not been pursued on a scale sufficient to cope with the problems. It has done little more than inflicting an insignificant dent into Java's problems and that too at great expense.[12] The ineffectiveness of the programme has been ascribed to many reasons, partly to poor organisation and administration, insufficient finance and failure to select suitable sites for settlements, some of which have been in areas where irrigation facilities were absent or would be impossible to provide. According to Mrs Hardjono :[13]

The problem is how to set about organising migration on a sufficiently large scale.

. . .It has been argued that for transmigration to make a significant contribution, it must move away from the traditional pattern in three important respects. It must :

(a) aim at ensuring an adequate supply of labour not merely for agricultural development, but in other sectors also for wider regional development;
(b) serve as a catalyst for regional development by using labour from Java to remove the obstacles to regional development and to mobility of labour arising from inadequate infrastructure; and
(c) offer migrants, both officially sponsored transmigrants and voluntary migrants, opportunities for employment as wage earners rather than as small holders.

During the Third Five Year Plan (1979-84), a large-scale expansion of the programme to conform to the above objectives is being contemplated. The target is to move some 100,000 families per year to transmigration settlements of about 5000 families each. The emphasis will be on regional development. It is recognised that many complex factors must be considered in this regard, including not only soil suitability, cropping patterns, cash crops, agro-industrial and other considerations, but also careful attention to site selection as potential growth centres with due regard to relevant infrastructure needs. When development takes place on these lines by establishing attractive growth centres, they will tend to attract completely independent settlers.* Over the past years, the government's transmigration efforts have provided some valuable experience leading to the awareness of the multifarious problems associated with transmigration. The policy departure from a limited programme of land development (with single objective) to regional development (with multiple objectives)

* According to Hardjono, this has in fact happened in Belitang in South Sumatra and Parigi in Central Sulawesi, as well as in the older colonisation area around Metro in Lampung.

marks a big change in the philosophy. The basic tasks of coordination, planning, carrying out and controlling supervision of development in transmigration areas, at both the national and regional levels still remain. While transmigration cannot solve all the population problems of Java and Bali, it can however make a significant contribution to development in a country fortunate in having land and water resources in abundance that have not yet been harnessed to the service of man. This kind of settlement and development of virgin areas calls for a total comprehension of development of an area along with an appropriate organisational, institutional and development administrative structure that would ensure that the numerous elements of development would fall in their places in appropriate sequences and in right time, thus ensuring unified and self-generating development in the area. In the absence of the critical minimum effort and investment needed to ensure development along these lines, however, the Indonesian programme of transmigration has remained a programme of limited achievements. Notwithstanding these limited achievements, it cannot be denied that transmigration has to be an important element in the Indonesian policy package, designed to bring about more equilibrating patterns of development within the country.

Local and regional development approaches

The planned and guided migration of population from the crowded to the relatively empty islands in Indonesia is only a partial solution to the problem of unequal development. In the already settled parts of the islands, the problem of persistent poverty among a sizable proportion of population and the problem of bringing about a more equitable and just distribution of the benefits of development to the target groups constitute the real problems. Indonesia has had a relatively long history of government intervention into rural development, but it was not until the late fifties and early sixties that, coupled with production incentives[14], specific policies were launched to improve the living conditions of the people in the rural areas.

Rural development policies in Indonesia have taken the form of development subsidies (unit or per capita grants) by the central government to the second (Kabupaten, Kotamadya) and lower levels (villages).* These are known as *Proyak Inpores* or Presidential instruction projects.

The village subsidy is a unit grant on flat basis introduced in 1969-70, when every village was given a development subsidy of Rs. 100,000. Since then, this subsidy has been increased upto Rs. 200,000 in 1976-77. The village subsidy has to be used for the rehabilitation and development of village infra-structures with voluntary labour and participation from the people. The cash is to buy materials; labour is to be supplied by the villages. The main emphasis of the programme is on encouraging diligence, self-help and cooperation among the villagers which forms the basis for the much-praised system known in the Indonesian language as *gotong royong* (literally meaning self-help). Under this scheme, the central government does not give the full budget but only a stimulus to attract voluntary labour in the villages.

The subsidy to the second level of the Government known as the *inpres Kabupaten—Kotamadya* was introduced in 1970-71 at Rs. 50 per capita. Subsequently it has been increased to Rs. 400 in 1976-77. This subsidy is intended as a catalyzer to create employment opportunities in the area by making appro-priate payment to labour in programmes of rehabilitation and infrastructures—roads, bridges, irrigation projects, market places and other local social-welfare projects.

Another development subsidy known as provincial develop-ment subsidy or *Provak Inpres Propinsi* was introduced in 1974-75. Alongside the provincial development subsidy, two more specific subsidies also exist, one for the construction of elemen-tary school buildings and another for health development consisting of poly-clinics, medicine, water supply, water closets (WC) and medical personnel. Lately market place subsidies and subsidies for replanting the critical areas have been introduced in the National Development Budget of 1976-77.

* Indonesia is a unitary State, divided into 27 provinces. Each province, which is also called the first level of government, is further sub-divided into Kabupaten (Regency or District) and Kotamadya (Municipality). The second level of government is the Kacamatan (Sub-regency or Sub-Dis-trict) and these are further divided into Desas, i.e. villages.

At present, rural development programmes of four categories may be distinguished at the sub-national levels :

(1) Projects planned and financed by the central government and implemented or performed by central government field officials of certain departments in the region (through the mechanism of deconcentration as it is usually called in Indonesia). Such projects are usually sectoral in nature.
(2) Projects planned and financed by local governments and performed by local officials (through the mechanism of decentralisation). These are identified as local projects.
(3) Projects planned and financed by the central government but performed by the local Government apparatus. (Through the mechanism of co-act in administration). The projects financed by the system of Presidential Grants-in-Aid (*Inpres*) projects come under this category.
(4) Projects planned and financed by the rural community through a process of community initiative (through the mechanism of community-initiated administration).

Experience in the implementation of the above categories or projects has brought out the lack of coordination and synchronization among the programmes which has inhibited the realization of their full development potential. As a result, the idea of preparing regional development plans by the provinces or in other planning units to synchronize these activities has emerged. Regional plans/studies have been undertaken by the Universities, by Provincial Governments as well as by a number of foreign consultants. A large number of such studies exist for different parts of Indonesia.

Since the Second Five Year Development Plan (1974-79), the development of villages or rural areas through regional development has been advocated and a new system of rural development has been introduced. In Indonesia, the strategy of rural development has been envisaged as a three-stage transformation of villages—from the stage of *Swadaya* village through *Swakarya* village to *Swasambada* and *Pancasila* village.*

The integrated approach to rural development administration in Indonesia which incorporates this village progression concept is called Local Development Working Unit. (UDKP's system is set up to manage the acceleration of the attainment of *Swasambada* village.) The main functions of UDKP are to set up priorities of rural development, to incorporate the sectoral, local, *Inpres* and rural community development programmes and to carry forward and to monitor the implementation of those programmes. The *Camat* who is the lowest deconcentrated Government official in the region is responsible for the implementation of the rural development programmes at the UDKP.

The basic principle underlying this innovated system of rural development is a principle of regionalisation of development planning, developed since the beginning of the implementation of the Second Five Year Plan in 1974. According to this concept, all the Development Centres throughout the country have been categorized as follows :

(a) main development areas with main development centres (4);
(b) large development areas with development centres (10); and
(c) medium development areas with development centres (85).

The above principle of regionalisation of development planning serves only as a mapping guide. It leads to the formulation of a development framework which may be called

* The *Swadaya* village is described as traditional village or society in which their economic conditions are only sufficient to fulfil their primary needs. The *Swakarya* village is described as a transitional village or society. These have been exposed to two external influences including the introduction of appropriate technology. The *Swasambada* village is described as a more advanced and modern village that will be further developed later as *Pancasila* village, which is the ultimate goal of the rural development programme. This (*Pancasila*) village can be described as a modern and advanced village within which all the basic needs of the people are to be fulfilled on a self-sufficiency basis. See "Integrated approach to Rural Development Administration in Indonesia" prepared by National Institute of Administration, Republic of Indonesia, 1978 (Mimeo).

the UDKP master plan. From the master plan, projects are drafted based on pre-determined programmes and formulation of the yearly budget for the projects. In reality, the implementation of the development plan is always performed on the basis of administrative territorial considerations, that is, national, provincial, Kabupaten (municipal) and Kacamatan apparatus with their villages.[15]

It may be noted in this context that what is obtaining in Indonesia today is a form of executive decentralisation and not political decentralization of decision-making.[16] Executive decentralization implies Central initiative and direction primarily through bureaucratic channels. At the present stage of economic development, this is obviously considered as more appropriate in terms of efficiency and timing of development efforts.

The Indonesian Economic Development policies are not, as yet, sufficiently informed by the principles of re-distribution with growth, anti-poverty strategies or structural reform of society. Land reform—from the start stigmatized as a policy of the communists—has been stopped, indeed reversed with the once redistributed land returned to former owners one way or the other or, passing into the hands of military officers.[17] Although the highly laudable principles of *gotong royong* (self-help) have been advocated, rarely does one find it working spontaneously. "People conditioned by more than 300 years of colonial administration wait on government action for everything".[18] Political decentralisation is far from sight. The current emphasis is on executive decentralization. The structural characteristics of the society are such that they effectively prevent the poor from obtaining access to many of the available development opportunities.

In Indonesia, the developmental spectrum varies from zero to saturation levels and closing the inter-regional gap would call for extraordinary effort, as it involves both initiating new resource developments as well as bringing about structural reforms and several adjustments and adaptations. The current policies aiming at redistribution of population on the one hand, and regional and local development on the other, are probably informed by rational approaches to meet these challenges. However, both in regard to the implementation of the transmigration programme as well as programmes of local and regional development, a major problem encountered is one of an unsound management system characterized by an ineffective institutional framework, scarcity of grass roots level leadership, inefficient coordination of development efforts, inadequate administrative machinery, lack of planning capabilities, weakness in implementation and lack of efficient control of the implementation activities of the development programmes. Both national as well as some internationally aided efforts are keenly aware of these problems.

Under these circumstances, the resulting processes of change from the existing policies have to be sufficiently prolonged, as they may contribute only slowly to the reduction of the various forms of inequalities characterising the Indonesian situation. We may not be able to hazard any spectacular progress in the situation for a long time to come and would therefore prefer to leave this account partly as a benchmark for a subsequent analysis.

NOTES AND REFERENCES

1. According to some critics, this is an over-simplification and marks significant variations in developments among the Outer Islands. North Sumatra, South Sulawesi and South Kalimantan, for instance, are much more densely settled than are the other parts of the Outer Islands. *Source :* Donald W. Fryer and James C. Jackson, *Indonesia*, Earnest Benn, London 1977.
2. Clifford Geertz, *Agricultural involution, the process of ecological change in Indonesia*, University of California Press, Berkely and Los Angels, 1963.
3. Shail Jain, Size distribution of income, a compilation of data (Washington D.C., IBRD 1975).
4. "Prospects for developing countries", World Bank, 1977.
5. Montek Ahluwalia, Nicholas G. Carter and Holis B. Chenery, Growth and poverty in developing countries, World Bank, Staff Working paper No. 309 (revised) May 1979.
6. M. Leiserson *et al*, "Employment and income distribution in Indonesia"—Studies in employment and rural development No. 51, Washington D.C. International Bank for Reconstruction and Development, 1978 (Mimeo).
7. Economic and Social Survey of Asia and the Pacific 1978, "Preliminary Draft", 15 Jan. 1979, E/ESCAP/L-26 Chap. III(C).
8. Hendra Esmara, Regional income disparities, *Bulletin of Indonesian Economic Studies*, Vol. XI, No. 1, March 1975.
9. Hendra Esmara, Regional Income Disparities, *BIES*, March, 1975.

10. Hendra Esmara, ibid.
11. Hendra Esmara, Regional Development and planning in Indonesia: A review of methodology, Paper to the Colloquium on Methods of Planning for Comprehensive Regional Development, United Nations Centre for Regional Development, Nagoya, 1976.
12. Donald W. Fryer and James C. Jackson, *Indonesia*, Ibid, p. 147.
13. J.M. Hardjono, *Transmigration in Indonesia*, Oxford Press, 1977.
14. Among the production incentive schemes to the farmers, special mention must be made of the several official attempts to deliver agro-inputs and credit to rice producers, the most important of which is a mass guidance programme called BIMAS and its various adapted, improved and renamed versions. There is evidence that these programmes have had some positive impact on yields, but have not yet been adequate enough to solve the 'rice problem'. For an assessment of the technical and social progress of paddy farmers, see David S. Gibbons *et al*, *Agricultural Modernization, Poverty and Inequality*, Published by Teakfield Limited, England, 1980, a study promoted by the Institute of Development Studies, University of Sussex, England.
15. See a report on Indonesia prepared by the National Institute of Administration of Jakarta included in the publication by United Nations. ESCAP on 'Local-level Planning for Integrated Rural Development"—Report of an Expert Group Meeting, Bangkok, 6-10 November, 1978.
16. Under the impact of the INPRES programme, one may perhaps conclude optimistically that the local government is altering from a holding operation to an agent of change. According to Irene Tinker and Millidge Walker, "Hard data is being collected, projections of growth and studies of projects are being made. Like anywhere also, mistakes are being made. Some studies are trivial, some data is falsified, some money is misused. What is important is that something is happening in the local areas with local input. It is not merely deconcentration of administrative rule-application, nor yet is it political decentralisation of decision-making; it is rather executive decentralisation"—Irene Tinker of Millidge Walker, Planning for Regional Development in Indonesia, *Asian Survey*, Vol. XIII, No. 2, December 1973.
17. IDS Mysore and IIED London: *Habitat Asia* : Volume 3, Indonesia by Donna Suri and R.P. Misra, 1978, (Mimeo) page 20.
18. I.D.S., Mysore and IIED, London (Ibid).

CHAPTER 9

UNDERDEVELOPMENT IN INDIA

THE CASE of India is particularly interesting in the study of the Geography of Underdevelopment because of the many facets of underdevelopment that it reveals and the numerous lessons and insights which its developmental experience affords. As a nation, India is wedded to democracy and individual freedom and follows a system of mixed economy in which direct control over the production of goods and services is imperfect. However, the country, under its federal political system, can draw upon a large armoury of instruments of State—persuasion, monetary and fiscal weapons, detailed controls, legislative changes and executive rules and institutional changes—to secure the aims and ends of planning. The Indian experience, therefore, is highly instructive in understanding the nature of various types of constraints on the effective implementation of the policies pursued within the framework of a mixed economy towards the accelerated development of underdeveloped areas and reduction in inter-regional disparities.

Inequalities in India, as in the other developing countries, exist rather sharply, both horizontally between populations living in different regions of the country and vertically between different segments of the population living in an area. The country, being vast—comprising an area of more than 3 million square kilometres of land—a multi-level analytical frame is required for analyzing and understanding the nature and extent of horizontal differentiation in development. Such a multi-level frame at the sub-national level may be afforded by the units like the State, the region[1] or the district. Indian researchers have employed these various area level aggregations in their analyses to bring out the different aspects of inter-regional disparities in the country.

The states are the federal units in the country having their own independent legislatures. There are twenty-two States and nine centrally administered territories in the country which are quite diverse and heterogeneous in character. The largest State, Uttar Pradesh is more than 29 times the area of the smallest (Tripura) and has 176 times the population of the smallest (Nagaland). The States also are diverse in their natural environments and resource distribution, in their populations, that are highly differentiated in terms of language, religion, caste, human institutions as well as in socio-economic levels of development. Notwithstanding this diversity and heterogeneity among the different states, any analysis of regional imbalances must start with States as the level of comparison, as they have most of the data on comparable basis.

Official estimates of State Domestic Product (SDP) exist and they may be considered as broad-based measures of well-being of the States' economy, notwithstanding their well-known limitations.[2] A set of comparable figures of per capita net state domestic product for three different periods for various states is given below to provide a broad-brush picture of inter-state disparities.

It may be seen from Table 9.1 that regional disparities in per capita net State Domestic Product have increased from 1.86 : 1 in 1962-65, 2.3 : 1 in 1967-70 to 2.7 : 1 in 1970-73. Also during 1970-73, the average per capita all India NDP stood at Rs. 655 at current prices. During the same period, the highest per capita SDP was recorded in Punjab (Rs. 1106) and the lowest was recorded in Manipur (Rs. 413) followed by Bihar (Rs. 439). There were 9 States with per capita SDP above the all India per capita NDP and as many as 11 States having less than the national average per capita NDP.

The ranking of the States according to the SDP does not change considerably over the three periods under consideration, even though there are some minor deviations. Punjab, Haryana, Maharashtra, West Bengal, Gujarat, Tamil Nadu, Himachal Pradesh are States which have higher than all India per capita income in all the three periods. Karnataka, Kerala, Andhra Pradesh, and Assam are States which have per capita income close to the all India average. Madhya Pradesh, Rajasthan, Uttar Pradesh, Orissa, Jammu and Kashmir and

TABLE 9.1 : **Average per capita net State domestic product at current prices (in rupees)**

States	Per capita net State Domestic Product		
	1962-65	1967-70	1970-73
Andhra Pradesh	386	537	645
Assam	393	581	610
Bihar	265	409	439
Gujarat	462	667	808
Haryana	445	810	974
Himachal Pradesh	n.a.	585	716
Jammu and Kashmir	302	426	524
Karnataka	373	552	703
Kerala	341	555	646
Madhya Pradesh	325	458	532
Maharashtra	478	686	799
Manipur	n.a	n.a	413
Orissa	306	488	566
Punjab	492	940	1106
Rajasthan	314	455	589
Tamil Nadu	400	558	675
Tripura	n.a	n.a	565
Uttar Pradesh	306	480	550
West Bengal	465	667	786
All India	372	556	655

Source : Quoted in Table 3 in Uma Datta Roy Choudhury and Grace Majumdar, Alternative Measures of Economic Development for Inter State Comparison : An exploratory study. Paper for Eleventh General Conference of Indian Association for Research in National Income and Wealth, Shillong 28-31 October 1977 (Mimeo).

Bihar are States with the per capita incomes lower than the all India average in all the three periods.

Using the State Domestic Product as the main criterion, the nature and magnitude of the problem of inter-State inequality in India has been investigated in a number of studies.[3] Sampath, adopting a measure similar to the coefficient of variation, studied the extent and change of inter-State inequalities in income in India during 1951-71. The conclusions that emerged out of his study were :

(i) There existed wide inter-State inequality in India at the beginning of the planning era and it declined steadily

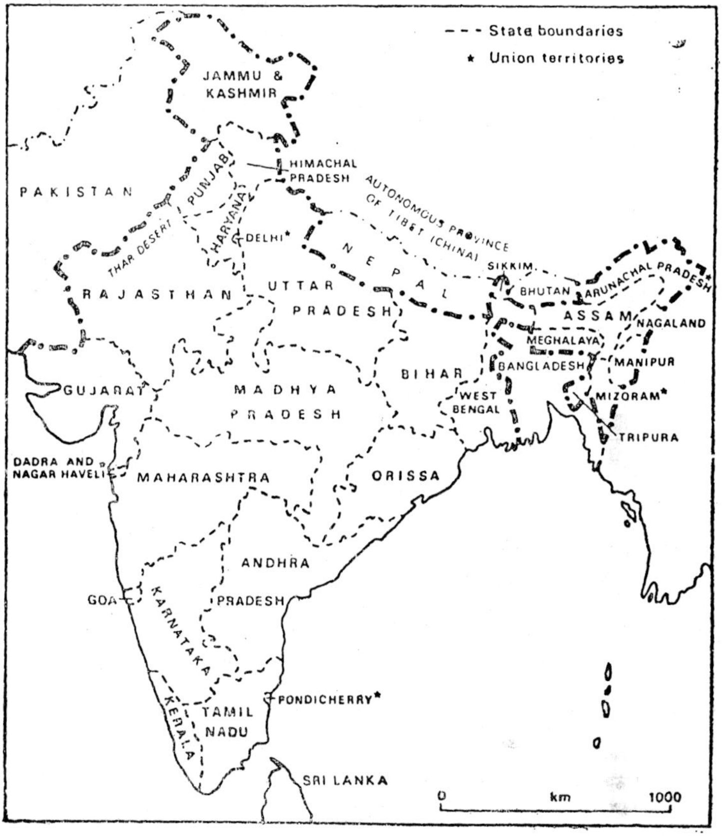

Fig. 6

until 1964-65 since when up to 1970-71, it increased steadily.

(ii) there existed wide inter-state inequality in the beginning in the distribution of net national non-agricultural income among the Indian States, but the trends are definitely towards reduction in such inequality which is evident from the fact that the share of non-agriculturally advanced States has been steadily declining and that of non-agriculturally backward States has been steadily increasing.[4]

(iii) in contrast to the trends in the non-agricultural sector, the growth policy of the agricultural sector during

1961-71 was such that it contributed significantly to the increase in the inter-State inequalities.*

(iv) there existed a positive correlation between regions' level of per capita industrial income and the degree of intra-State inequality in per capita income over districts in 1955-56, implying that the level of intra-State inequality depended highly on its level of industrialisation.

The manner in which regional disparities measured in terms of State Domestic Product have changed during the process of economic growth leading to convergence or divergence has been specially investigated in some studies. Ashok Mathur's study merits special attention in this context.[5] Using a comparable time-series data for a period of 25 years (1950-51 to 1975-76) and adopting primarily the coefficient of variation method aided by different statistical tests to check various hypotheses, the author reaches the conclusion that up to the early sixties, there was a downward, though fluctuating movement of regional income inequalities, but since the middle sixties, there has been a sustained increase in inequalities, a point which is also corroborated by Sampath in the study referred to above. Thus the 25 year period of planning in India appears to be marked by a broad tendency towards convergence in the first half and divergence in the latter half, suggesting a U-shaped curve of regional inequalities, and not an inverted U-curve as in Williamson's hypothesis.**

* The analysis revealed that the reduction in overall inequality over the period 1960-61 to 1964-65 was entirely contributed by the non-agricultural sector and the increase in inter-regional inequality over the period 1964-65 to 1970-71 was entirely due to the agricultural sector. This was possibly due to the selective agricultural development policies (IADP, HYV etc.) leading only to a few states, (for example, Punjab, Haryana and Andhra Pradesh) recording very high growth rates in agriculture.

** According to Ashok Mathur, the apparent contradiction between the two may not necessarily imply a refutation of the Williamson hypothesis. If the basic underlying functional relationship between per capita income and regional inequalities is seen as a long term relationship, evident only from trend data spanning over half a century or more, the Indian experience could be interpreted as a phase of cyclical movement (fluctuations during medium term time spans) within the overall trend curve of Williamson.

The level of the domestic product is merely an indicator of the productive capacity of a country or region, but it fails to give us an idea of the proportion of expenditure which goes into personal consumption and the way the expenditure is distributed among the various items of consumption. An exploratory study of the inter-State differentials in the levels of living was attempted by Ganguli and Gupta[6] using certain economic and demographic variables with special reference to the trends in the inter-State disparities during 1955-65. In some cases, where data were available, the analysis was extended to cover the year 1970 as well. Rank correlations were used to investigate the inter-relationships between the economic and demographic variables and the various composite indices of the levels of living. We may quote the main conclusions of the study here.

> While the levels of per capita domestic product and the levels of living have a close relationship, it is tenuous when the levels of per capita consumption expenditure are considered. It is found that the states with high levels of living did not show high rates of growth of the domestic product. Widening inter-State disparities in the levels of domestic product were found to be associated with the fact that the States seemed to be coming closer in respect of their levels of living. States with high proportions of their population engaged in primary activities usually showed low levels of living and hardly any significant relationship was observed between the tertiary sector and the conditions of living...... As regards the literacy factor, the States with high literacy levels have shown better levels of living.[7]

On matters of policy at the national level, the inter-State inequalities in income and possible levels of living too are consciously taken note of in decision-making by the Government, particularly on crucial allocative decisions, which are supposed to have equilibrating effects on the economic performance of the States. Thus both in the criteria adopted by the Finance Commission* for the devolution of tax revenues to the

* Every five years, the Government of India appoints a Finance Commission to go into the various aspects of States' finances and to recommend (*Contd.*)

States and in the formula employed for allocation of central assistance of plan finances to States* 'the inverse of State income per capita' figures as one of the guiding considerations. With the States becoming increasingly articulating and conscious of their developmental lags, the problem of regional disparities can no longer be dispensed with at the national level as a long term objective of planning. The Five Year Plan of 1978-83 is conscious of this fact and has several in-built features, rendering it increasingly biased in favour of the backward states. We shall discuss this aspect further in the sequel.

While inter-State disparities in development are baffling in themselves, in a big country like India, the intra-State disparities (or inter-district disparities) reveal the greater complexity of the problem of regional imbalances, which merit consideration not merely at the State level, but also at the national level, as these have often political overtones. Thus the spatial configuration of underdevelopment gains greater depth and clarity when analysed in terms of districts. A broad picture of the spectrum of inter-district disparities in the country may be obtained by scanning through a table assembled by Gulati[8] (see Appendix) which sets out the dimensions of inter-district disparities in a vector of thirty-two variables varying over 336 districts, wherein some conventional measures for reflecting the range of disparities such as the Mean, Standard Deviation, coefficient of variation and skewness are given. A perusal of this table shows that the inter-district variations are substantial.

the principles and method of devolution of central revenues to the State Governments. The recommendations of the Finance Commission are statutory and binding on the Government of India.

* Since the Fourth Five Year Plan, the quantum of central assistance (Plan resources) to States is being governed by a formula known as the Gadgil formula (named after Prof. Gadgil who was the Deputy Chairman of the Planning Commission and who was instrumental in evolving this formula). The formula gives weightage in central plan allocations to population of the State (60%), States below national average of per capita income (10%), Tax effort in relation to per capita income (10%), special problems of the State (10%) and outlays required for major irrigation and power projects in the State (10%). More recently (March 1979) the criterion of 'inverse of State income per capita' has been introduced for giving greater weightage to backwardness of States.

Studies attempting to portray the levels of development among the districts through selective indicators have been conducted by several authors which vary in the choice of indicators as well as in methodological refinements introduced for combining a large number of disparate indicators. (Schwartzberg 1962; Asok Mitra 1961; Prakasa Rao, Sundaram et al 1971; Biplab Das Gupta and Morris Jones 1975; Pal 1975[9]). More recently (1978), this author conducted a study in the Planning Commission, using the data of the seventies, and delineated the less developed districts in India on objective criteria. The study used 14 variables and the Principal Components Analysis.* The results of the study (see Map) indicate that the relatively less developed districts of India are in the east, centre and middle south parts rather than in its north, west and deep south. The contiguous areas of medium and high level development seem to fall in five areas (i) in the west and south, extending from Gujarat down through Maharashtra which is further linked up with the southern districts of Karnataka, Kerala and the whole of Tamil Nadu in the south; (ii) in the north-west, including Punjab, Haryana and Delhi, western U.P. and parts of Rajasthan (Ganganagar and Bikaner districts); (iii) in the east, the area extending from Calcutta through the Damodar Valley Industrial Belt; (īv) Mahanadi Delta area in Orissa, and (v) the Krishna-Godavari deltas in Andhra Pradesh.

The relatively less developed areas which we may characterize as "Underdeveloped India" (categories 'low' and 'very low' in Fig. 7) cover a vast stretch of the eastern, central and middle south India (excluding the areas mentioned above). Within this area, stand out small areas around the urban nodes of Ajmer, Kota, Gwalior, Ujjain, Bhopal, Agra, Kanpur, Lucknow, Allahabad, Varanasi, Patna, Rourkela, Jabalpur etc. like "islands," representing a spotty kind of development that has not had its spread effects in the rural areas. The areas in the very lowest levels of development appear to be (i) the districts of Bastar, Sidhi, Surguja, Panna, Tikamgarh and Jhabua

* The study (not yet published) was undertaken as part of the work of the Committee on Backward Areas Development, constituted by the Planning Commission. The variables used in the study are listed in Appendix.

in Madhya Pradesh, (ii) Koraput and Kalahandi in Orissa, (iii) Palamau and Purnea in Bihar, (iv) Jaisalmer, Barmer, Jalore, Jodhpur, Dungarpur and Banswara in Rajasthan, and (v) Punch, Riasi and Udhampur in Jammu and Kashmir.

The geographical configuration of underdeveloped areas revealed in the above analysis and shown in Fig. 7, clearly indicates that the backward areas of India are quite extensive in character falling in diverse regions of markedly different natural resource endowments and socio-economic environments. In terms of this diversity, the "Underdeveloped India", appears to be as diverse as the sub-continent itself. Om Mathur's characterization of the Indian regional problem seems appropriate. "The regional problem in India is complex. It is not the typical north-south problem of Italy, but one in which the disparity problem is interspersed throughout the country in a very asymmetrical manner so that poverty areas exist within each region...there are regions that are rich because of their raw material endowment; there are regions that are poor despite their rich resources; and finally poor regions because of their poor resources.[10] Our aim in emphasizing this point at this stage is to highlight the fact that from the policy angle, Underdeveloped India does not constitute a single problem region. In terms of spatial relations and accessibility to core regions and modern developments, pressure of population on land, resource endowments, level of technology and skills of population, socio-cultural traits of the population, and historical factors in development, a taxonomic breakdown of the underdeveloped areas into segments having broadly similar problems and needs can be attempted. In our opinion, the problem areas identified and classified according to the nature, cause and severity of backwardness are the best type of geographical units for the application of regional development policy. Studies conducted in the Indian Planning Commission have revealed that the backward areas of the country may be grouped into five fairly compact regions, each with its own complex of symptoms, causes and remedies. These are :

1. The North east region of India;
2. The tribal belt in East-central India;

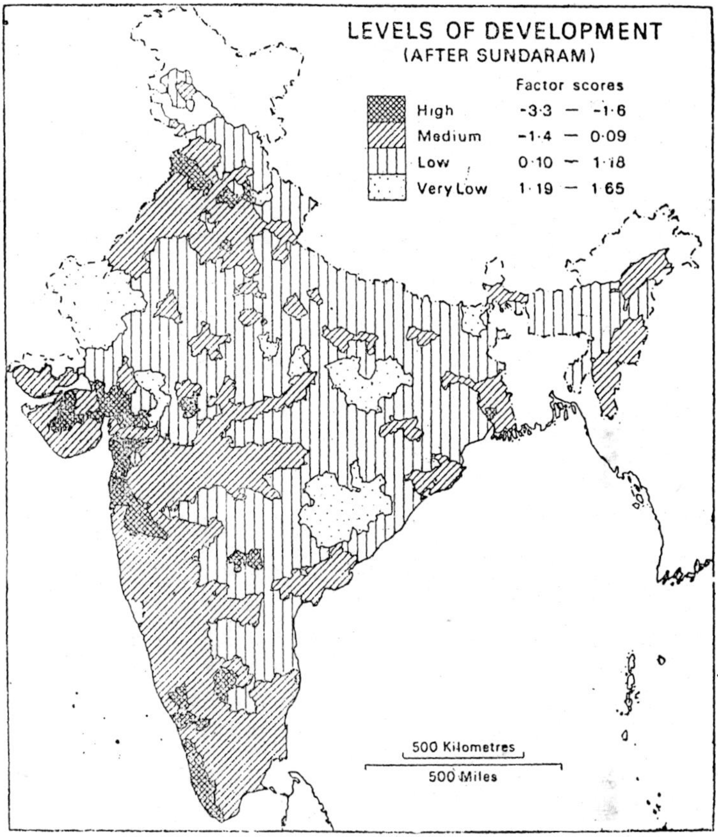

FIG. 7

3. Eastern Uttar Pradesh and North Bihar;
4. Bundelkhand region of Uttar Pradesh and Madhya Pradesh; and
5. Ecologically unbalanced areas with special problems, for example, drought prone areas, desert areas, ravine areas, flood prone areas, hill areas, and so on.

The relevance of the above classification for policy treatment will be discussed at a later stage in our discussion.

A broad picture of the spatial inequalities (with States and districts as the unit areas of analysis) has been presented in the foregoing paragraphs. We may now supplement this with an account of inter-personal inequalities among the population. This is important because in developing countries, there is a close nexus between the two.[11]

In India, poverty is known to be extensive, although with the exception of a number of good papers contained in Srinivasan and Bardhan's *Poverty and income distribution in India*,[12] not much work appears to have been done to describe its spatial incidence with precision and refinement. Empirical studies, more at a global level, setting out the overall dimensions of the poverty problem—its extent and intensity at the national level—exist. Mention must be made of the studies by Ojha,[13] da Costa[14] and Dandekar and Rath.[15] They have, by and large, helped in defining the poverty line in statistical terms.

The criterion of poverty in India is the absolute minimum level of consumer expenditure required to secure a diet adequate at least in respect of calories.[16] In the Draft Plan 1978-83, the Planning Commission has defined the poverty line on the basis of recommended nutritional requirements of 2,400 calories per person per day for rural areas and 2,100 calories per person per day for urban areas.* In rupees, the poverty line is the midpoint of the expenditure class (in 1973-74 consumer expenditure data) in which the calorie needs are satisfied. Thus the poverty line turns out to be Rs. 61.8 and Rs. 71.3 for rural and urban areas respectively at 1976-77 prices. According to the Planning Commission, the percentage of people living below the poverty line in 1977-78 was 47.85 per cent in the rural areas and 40.71 per cent in the urban areas (All India 46.33 per cent). In terms of numbers, this would mean about 250 million people.

The above overview of poverty in arithmetic terms[17] no doubt enables us to appreciate the magnitude of the problem. It has also served to provide a bench-mark for future comparisons. But for planning purposes and for greater specificity in programme-decision-making, one ought to have a clear picture of the spatial variations in the levels of poverty and relate the

* This is based on the recommendations of the Nutrition Expert Group (1968) who have estimated the weighted average daily calorie per capita norms at 2,435 for rural areas and 2,095 for urban areas. When rounded off, this yields the figures 2,400 for rural areas and 2,100 for urban areas.

same to the various components of the socio-economic system. The national aggregates of poverty estimates were made from consumer expenditure data of National Sample Survey for 1973-74. When attempting to use the same data for disaggregations at sub-national levels, one comes across serious difficulties, both statistical and conceptual. Firstly, while making such estimates, area-wise, it is necessary to take into consideration, differences in respect of sex, age, occupation, race, geographical conditions etc. since calorie requirements for a minimum level of living depend on all these factors. Secondly, the consumption basket also differs from area to area, depending upon local tastes, availability and so on. Thirdly, the price differences among areas introduce yet another complexity, as they constitute an important determinant of the poverty norm. All these render such an exercise almost impossible. In the face of such methodological problems, a somewhat crude approach would be to apply an All-India poverty norm to the State level consumption data and estimate the State-wise percentage of population below the poverty line. Such estimates, crude though they are serve to bring out the relative differences among States in regard to the concentration of the poor. An estimate of this kind is presented in Table 9.2.

This table shows that the two States of U. P. and Bihar have in them nearly 29 per cent of India's poor. The States of West Bengal, Tamil Nadu, Madhya Pradesh, Maharashtra, Andhra Pradesh, Orissa and Karnataka have proportions varying from 5 per cent to 10 per cent. In Kerala, Gujarat, Rajasthan, Assam, Haryana and Punjab, their proportions to All India total varies from 1 to 5 per cent. In other States their proportion is not very significant.

The picture of incidence of poverty by States, however, marks significant local and regional variations. A far more meaningful pattern may be obtained if this picture can be disaggregated further region or district-wise. However, on account of the methodological difficulties enumerated above, such disaggregation cannot be attempted with precision. In the alternative, we may perhaps turn our attention to a set of relatively small number of independent poverty-related variables, to provide by inference and indirect evidence what direct measures of poverty may fail to provide. We consider three poverty-

TABLE 9.2 : **Poverty percentage among States**

States	Poverty Percentage	Percentage of population	
		Rural	Urban
1	2	3	4
1. Andhra Pradesh	7.39	61.57	54.31
2. Assam	1.71	50.98	49.64
3. Bihar	12.82	49.18	47.47
4. Gujarat	4.26	50.44	44.02
5. Haryana	1.15	32.93	46.03
6. Himachal Pradesh	0.42	25.94	19.92
7. Jammu and Kashmir	0.52	50.44	66.01
8. Karnataka	5.15	58.16	50.56
9. Kerala	4.72	55.09	58.57
10. Madhya Pradesh	8.51	61.09	51.41
11. Maharashtra	7.88	55.36	40.90
12. Manipur	0.12	46.64	34.99
13. Meghalaya	0.12	33.73	11.92
14. Nagaland	0.07	N.A.	8.15
15. Orissa	5.26	69.88	53.16
16. Punjab	1.04	26.61	53.31
17. Rajasthan	3.96	43.04	47.36
18. Sikkim	0.02	N.A.	N.A.
19. Tamil Nadu	8.88	63.87	59.85
20. Tripura	0.37	54.41	29.40
21. Uttar Pradesh	15.81	56.85	59.26
22. West Bengal	9.80	64.77	36.92
Total	100.00		

Source : For col. 2, Table A. 2 in the "Note of dissent by Prof. Raj Krishna entitled : A more equitable distribution of resources" in the Report of the Seventh Finance Commission, Govt. of India, 1978.
For col. 3 Planning Commission.

related variables, as particularly significant in the prevalence and persistence of poverty in the Indian context. These are :

(i) Concentration of landless agricultural labourers;
(ii) Degree of unemployment; and
(iii) Consumption standards.

Dandekar and Rath, in their classic study[18] have related the distribution of agricultural labour households to annual per capita consumer expenditure and have shown that the rural

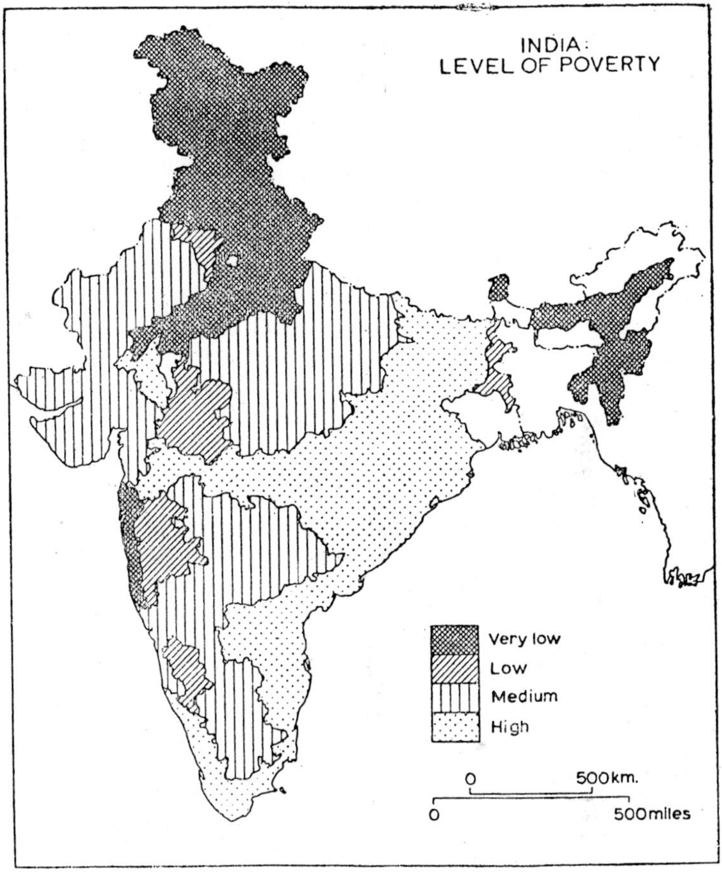

FIG. 8

poverty at the lowest level is largely accounted for by the agricultural labour households. According to Mellor, the indigence of the landless labourer constitutes "the most intractable component of rural poverty".[19]

Hence the concentration of agricultural labourers in an area must be regarded as a measure of rural poverty. The second variable the degree of unemployment is chosen here because all national and international studies in developing countries in recent years have revealed that poverty in these countries is closely associated with the lack of socially productive employment opportunities. This is the reason why the socio-economic objectives concerned with employment creation and distribution are increasingly becoming important in the planning policies of developing countries. The reasoning is simply that an increase in the level of productive employment would contribute to more equal distribution of income and improvements in living standards for the majority of the population. The analysis made by Dandekar and Rath[20] has revealed that the chief characteristic of the rural agricultural population of India is not that it is acutely unemployed or under-employed during the slack agricultural season, but that it is irregularly and inadequately and unemployed intermittently throughout the year. Hence we have used the average rate of unemployment for the year as a whole (rather than the peak rate during the year) to reflect this endemic rather than seasonal unemployment. The third variable used by us is consumption standards, which gives a more direct measure of the well-being or standard of living of a people.[21]

For the purpose of this analysis, the regions delineated by the National Sample Survey (NSS) were used and where necessary, the data were transformed to conform to these regions. The statistical distribution in the case of each variable was then classified according to intensity and then a composite picture was derived. This composite picture shows the levels of poverty and may be seen in Fig. 8.

The map of poverty levels by and large conforms to the map of levels of development, but there are striking differences in at least two areas to which we shall presently draw attention. The areas of acute poverty—the most distressed areas—are spread all along the East Coast of India fanning out in the north to include the whole of Bihar and also extending as a belt westwards across the belly of India, (including Eastern Madhya Pradesh, Inland Northern, Inland Eastern and Eastern Regions of Maharashtra). The belt of acute poverty also extends along the west coast to include the whole of Kerala. The relatively low levels of poverty are to be found in three belts : (i) in the north and north-west stretching across J & K, Punjab and parts of Haryana and U.P.; (ii) in the west—across Gujarat and coastal Maharashtra; and (iii) in the north east including all the north eastern states of the country.

The two major anomalies revealed in this picture are those presented by the Kerala coast and the north-eastern region of the country. Whereas the former stands out as an acute poverty area, the latter shows up as a no poverty area. Alice Thorner[22] has, in a recent study, investigated the strange case of poverty revealed by Kerala and has drawn attention to the findings contained in a report prepared by the Centre for Development Studies (CDS) in Trivandrum for the United Nations[23]. The Food Balance Sheets compiled by the CDS suggest that the NSS has severely under-reported certain items particularly important in Kerala dietary habits[24] revealing a discrepancy between the 1620 calories of the NSS estimate for 1961-62 and the 2337 calories availability shown by the CDS food balance sheets for the same year. Thorner has also examined various other economic indicators[25] and concluded that they do not build up a consistent picture of extreme poverty in Kerala. In the case of the north-east region as well as the hill and tribal areas of the country, the unemployment and consumer expenditure data assembled by the NSS do not reveal acute poverty conditions. Perhaps these areas represent the typical low level equilibrium trap of underdevelopment frequently mentioned in development literature[26]. On the whole, the map of poverty levels correlated with the map of levels of development and interpreted with caution and understanding provides a fairly meaningful mosaic pattern of the spatial distribution of underdevelopment in the country. With the finer spatial breakdown of some critical data, a more detailed picture could be built up. At this level of statistical refinement, however, this is not possible.

TABLE 9.3 : **India Inter-district disparities**

Variables	Mean	S.D.	C.VX109	MIN	MAX	SKEWNESS
	(1)	(2)	(3)	(4)	(5)	(6)
Demographic						
1 D/SQM	922.2	4696.7	509.2	2.0	73642.0	13.0
2 SR	937.1	82.7	8.8	617.0	1237.0	—1.0
3 R/1000P	839.4	152.1	181.2	0.0	1000.0	—2.8
4 GR/100	23.3	11.4	48.9	0.2	105.2	2.4
5 IMMG/100	12.7	10.1	79.5	1.3	83.0	2.7
6 C.W.R.	73.4	9.2	12.5	48.3	99.1	0.0
7 MAN	21.1	2.6	12.3	15.0	27.5	0.0
8 FAM	15.9	2.0	12.6	7.5	22.8	0.6

Overall Socio-Economic						
	(1)	(2)	(3)	(4)	(5)	(6)
9 LITCY	221.3	101.6	45.9	45.0	623.0	1.2
10 PCI	227.2	101.1	44.5	84.0	1510.0	7.0
11 APTI	50.8	14.8	29.1	0.0	84.1	—.6
12 MPTI	14.8	9.2	62.2	0.05	57.6	2.0
13 SPTI	34.0	11.0	32.4	8.9	94.1	0.9
Agricultural						
14. ADC/ 100 NAS	27.4	75.5	275.5	0.1	74.5	6.7
15 GAI/100GAS	19.3	21.5	111.4	0.1	74.5	4.1
16 HH 0-5/CHH	50.7	25.5	50.3	0.9	100.0	—.1
17 THL/CHH	9.8	16.4	167.3	0.01	46.7	6.9
18 HDWR/ 100W	11.0	12.8	116.3	0.1	90.0	2.8
19 C & A/ 100 NAS	46.4	26.6	57.3	9.0	214.0	1.6
20 C&A/100 RW	77.3	16.3	21.1	0.8	97.8	—2.4
Participation Rates in Traditional Sector						
21 MPR	57.8	4.8	8.3	41.4	81.0	0.1
22 FPR	31.3	16.0	51.1	2.2	69.2	—.0
23 MA/100 WM	66.4	17.1	25.8	0.02	90.5	—1.5
24 S & FW/100 NAW	57.9	16.0	27.6	5.5	91.8	—.9
25 WHI/100 WP	6.5	5.6	16.2	0.02	82.8	7.9
Participation Rates and Infrastructure in Distributive Trade & Manufacturing						
26 WRT/1000P	13.7	7.3	53.3	1.0	65.0	2.2
27 WMG/1000P	13.9	17.1	123.0	0.0	160.0	3.1
28 CHUBH/ 1000 H	3.4	2.9	85.3	1.0	28.0	4.5
29 CHUMI/ 1000 H	1.4	1.4	100.0	1.0	23.0	11.8
30 SRD/ 100 SM	182.3	221.6	116.1	0.0	1760.8	4.3
31 ERE/100 IB	5.7	7.3	128.1	0.0	52.2	2.5
32 WRF100 /W	1.6	3.1	193.7	0.0	30.0	5.1

List of Variables

Number (1)	Abbreviated Name (2)	Variable (3)
1	D/SOM	Persons per sq. mile
2	SR	Females per 1,000 males

(1)	(2)	(3)
3	R/1000P	Rural population per 1,000 of total population
4	GR/100	Percentage increase of population, 1951-61
5	IMMG/100	Immigrants per cent of total population, 1961
6	CWR	Children 0-4 per cent of women aged 15-44, 1961
7	MAM	Male age at marriage
8	FAM	Female age at marriage
9	LTTCY	Crude literacy rate, 1961
10	PCI	Per capita income
11	APTI	Per cent of income from agriculture
12	MPTI	Per cent of income from manufacturing
13	SPTI	Per cent of income from service industries
14	ADC/100NAS	Area under double crop per cent of net area sown
15	GAI/100 GAS	Gross area irrigated per cent of gross area sown
16	HHO-5/CHH	Households cultivating (0-5) acres per cent of all cultivating households
17	THL/CHH	Pure tenancy holdings per cent of all cultivating households
18	HDWR/100W	Hired attached workers per cent of all workers at cultivation
19	C & A/100 NAS	Cultivators and agricultural labourers per 100 acres of net area sown
20	C & A/100 RW	Cultivators and agricultural labourers per cent of rural working population
21	MPR	Male participation rate
22	FPR	Female participation rate
23	MA/100 WM	Males working in agriculture per cent of male working population
24	S & FW/100 NAW	Single & Family Workers in non-agriculture per cent of total non-agricultural population
25	WHI/100 WP	Workers in household industry per cent of total working population
26	WRT/1000 P	Workers in retail trade per 1,000 of total population
27	WMG/1000 P	Workers in manufacturing per 1,000 of total population
28	CHUBH/1000 H	Census houses used as business houses or offices per 1,000 houses
29	CHUMI/1000 H	Census houses used as medical institutions per 1,000 houses

(1)	(2)	(3)
30	SRD/1000 SM	Males of surfaced roads per 1,000 sq. miles of area
31	ERE/100 IE	Establishments run on electricity per cent of all industrial establishments
32	WRF/100 W	Workers in registered factories per cent of all workers.

Source : S.C. Gulati, Dimensions of Inter-district disparities, *Indian Journal of Regional Science*, Vol IX, No. 2, 1977.

Programmes and policies for the accelerated development of the underdeveloped areas

Having surveyed the developmental and poverty levels in the country, we may now turn to the programmatic and policy dimensions of the problem of underdevelopment. Physical and socio-cultural diversities and demographic variations characterising Underdeveloped India make it clear and imperative that the "underdevelopedness" in the Indian context cannot be tackled through devising a simple formula or a general scheme, but needs an approach that is increasingly area specific, people specific and problem specific. Accepting this as a valid proposition for the development of backward areas, it is necessary at this stage to inquire to what extent India's programmes and policies are informed by such an approach. This will require a close examination of the specific policies, strategies and programmes pursued at the national and sub-national levels for the accelerated development of the more underdeveloped areas of the country.

At the outset, we may note that the concern in India for the development of the relatively more underdeveloped areas of the country has stemmed from an increasing perception by the central government of the problem of growing inter-regional disparities, the reduction of which has been one of the prime national policy objectives in planning. Although the successive Five Year Plans have been laying varying emphasis on this aspect of the problem, it was largely rhetorical and platitudinous until the Fourth Five Year Plan[27]. The Fourth Plan was instrumental in making certain far-reaching changes such as formulating the principles of allocation of central assistance to the States under which the relative backwardness of a State was given specific weightage.* The Fourth Plan had also the distinction in pioneering a systematic effort towards the accelerated development of backward areas by initiating certain specific central programmes calculated to tackle this problem on a selective scale. Thus industrial backwardness was specifically sought to be tackled through selective fiscal and incentive programmes. Similarly certain target groups like the small and marginal farmers and the tribal people were covered under special central programmes. Also for the drought prone areas in the country and certain category of hill areas, special centrally assisted programmes were evolved. In relation to the backward areas, the Fourth Plan took a comprehensive view of the factors of backwardness and proposed that a multi-directional Area Development approach should be adopted in order to accelerate the development of backward areas. Since each backward area represented a unique combination of factors, it was recognised that no uniform programme could be successfully conceived and imposed from the national level. It was, therefore, felt that an essential pre-requisite for accelerated development was the evolution of appropriate location-specific strategies based on a careful identification of the causes of backwardness as well as the potential available for development.[28] There are significant statements and it must be truly said that the Fourth Plan initiated the thinking on the Area Development approach and also made concrete the catalytic intervention and the supplementing role of the centre in developing the underdeveloped areas located in the various States. The Fourth Plan is therefore an innovating plan in many ways and must indeed be seen as a watershed in Indian planning, where one finds the departure from a mere growth-oriented approach to one characterised by the principles of growth with equity.

* During the Fourth Plan, a formula for guiding the allocation of central assistance to the States was evolved, known as Gadgil formula. Under this formula weightage is given to various factors as follows: (a) population 60 per cent (b) per capita income less than the national average 10 per cent (c) for the continuing major irrigation and power schemes in a State 10 per cent (d) tax effort made by the State 10 per cent and (e) special problems of the state. Weightage 10 per cent for backwardness in this formula is contained in criteria (b) and (e).

During the Fifth Plan, the development of backward areas assumed an even greater importance in view of the emphasis laid on raising the level of living of the bottom 30 per cent of the population, a significant proportion of which are found in the backward areas. The area programmes were multiplied and they included programmes following the resource-based or problem-based development approach, the target group approach, the incentive approach or the comprehensive area planning approach. The classification is as follows.[29]

I.	Resource-based/problem-based area programmes	— The Drought-prone Area Programme, the Command Area Development Programme, Hill Area Development.
II.	Target group programmes	— The SFDA Programmes; The Tribal Development Agency Projects.
III.	Area specific incentive programmes	— Concessional Finance; Investment Subsidy and Transport Subsidy Schemes.
IV.	Comprehensive Area Development Programmes	— Sub-Plans for the Hill and Tribal Areas.

The areas covered under the major programmes mentioned above may be seen in Fig. 9. The fact that as many as 340 out of the 398 districts in the country have been covered by one or more of these special programmes is an indication of the range of central intervention in the sphere of area development.[30] This plethora of area programmes are now at the crossroads, undergoing rapid adjustments to be integrated into an operable national approach to planning at the micro-level. We shall come to this development presently.

Aside from area planning, another direction in which the acceleration of the development of backward areas and the reduction of regional imbalances in the country were sought to be tackled throughout the successive plan periods has been through the

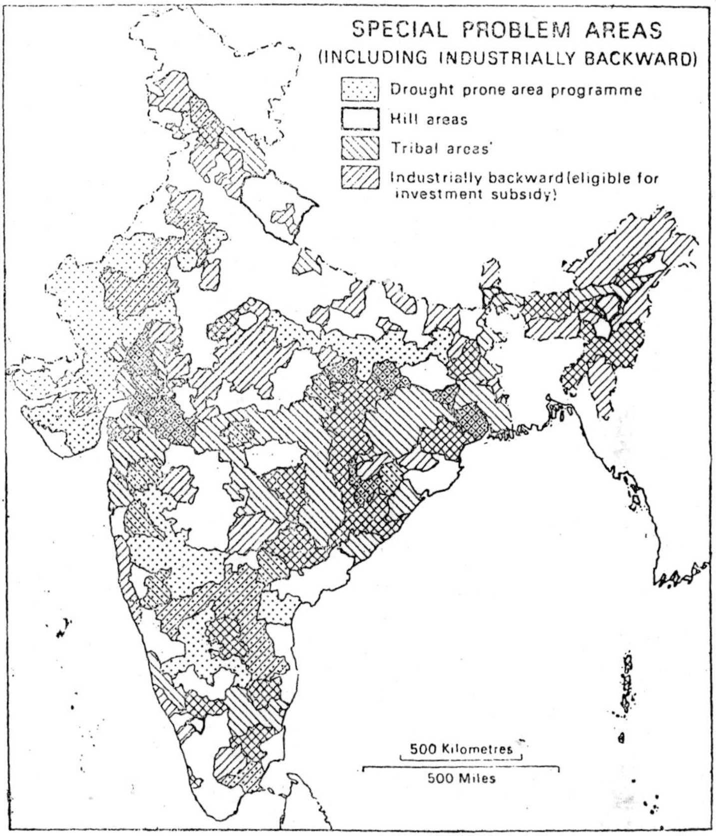

FIG. 9

emphasis on the subnational effort in district planning. In the late 1960s, when growth with social justice became the predominant objective, district planning[31] was advocated in recognition of the idea that without district or local planning, there could be no hope for acceleration of the development of the underdeveloped areas as well as regional closure. However, for various reasons, District Planning did not match up to these expectations sufficiently to become a major force in intermediate and local level planning. The disillusionment with district planning on the one hand and the change in the government on the other with its accompanying policy changes as well as the distinctly global trends (which is also parallelled in India by similar trends) away from gross increases in production and toward more distributive justice, have led to shift to the small area-based approach in planning that is grass roots level planning with the rural poor as the focus and the development block (which is a sub-district unit) as the basic unit for planning, pointing to a system of decentralised action as well as decision-making informed by social justice.[32] All the ongoing programmes in the country (including the specific backward area programmes which were mentioned earlier) are now getting adjusted to this new strategy and approach during the current plan period, that is, 1978-83).

The above account of contemporary trends in sub-national level planning in India was introduced to provide a backdrop to our discussions. More relevant to our theme are the programmatic devices and policy approaches applied to the backward areas problem in the country, the experiences gathered and the lessons learnt therein. These are critically examined in the sequel. At this stage, we may note that the Indian experience is characterised by variety, both in terms of situation as well as on methodology bearing on solutions and may have important lessons for many other developing countries.

Among the numerous specific approaches tried by the Indian Government, the following ten policy measures may be specially mentioned :

(i) Capital and technology transfers to backward areas;
(ii) incentive policies for peripheral areas;
(iii) growth centre policies;
(iv) the basic needs-oriented approach;
(v) target-group oriented approach;
(vi) resource problem-oriented approach;
(vii) asset re-distribution policies;
(viii) sub-plan approach e.g. hill and tribal areas;
(ix) regional planning approach for special problem areas; and
(x) integrated rural development oriented to full employment in Development Blocks.

In the following sections, an attempt will be made to pull together available empirical evidence relating to the performance of the above regional/area programmes and policies.

References to case studies will also be made wherever such studies are available.

Capital and technology transfers

As in many other developing countries, in India too, capital and technology transfers have been used where feasible, as one of the major instruments of public policy to stimulate the development of backward areas. In a mixed economy like India, the public sector operates many key or cornerstone enterprises. Industrial policy in India is designed to permit the use of such instruments in the location of industries in industrially lagging areas, "provided the location is otherwise suitable"[33] that is subject to their techno-economic feasibility and viability. In the licensing of private sector industries too, the development of backward areas is a guiding consideration. Thus in the location of some of the heavy industrial complexes in the north-eastern part of peninsular India (which is particularly rich in mineral raw materials) for example, Bhilai, Rourkela, Ranchi, Bailadila and Bokaro, the backwardness of the rural areas and the need for locating a propulsive industry capable of generating economic and social change in the region were not insignificant considerations. But the way developments have taken place show that there have not been much spread-effects and such developments stand out as islands amidst a vast rural area characterised by serious backwash effects. Most of the output of these public sector industries comprises products in the first stage of processing, which are exported after this stage, to feed the industrial activities and their concomitant multiplier effects are elsewhere. Thus there has been no regional development around these capital intensive developments. Not only have the developments been distorted and spatially disarticulated, but also they have been positively disruptive of the traditional occupation and economic and social life of the tribal communities living in their immediate neighbourhood. (Sundaram and Misra 1978; Sundaram 1977; Chattopadhyaya and Raza 1975; Roy Burman 1970; Sen Gupta 1976).[34] Thus the development of large, capital-intensive industries using high technology introduced in backward areas have tended to bypass the tribal communities of the area. According to some, they have even tended to push the tribal communities back on the economic scale. Experience in India has clearly shown that policy measures of this kind must not be pursued in isolation, but must be accompanied by (a) protective measures concerning themselves with the softer spots of the socio-economic scene and (b) positive action programmes which will enable the community to become active partners in the new development processes (Sharma 1978).[35]

Incentive policies for peripheral areas

Among the incentive policies, special mention must be made of the assistance provided to the small entrepreneur in starting industries in the backward areas. On the basis of the recommendations made by a high-level committee,* two incentive schemes applicable to industrially backward areas in the country are operating since 1970. One of them, called the scheme of concessional finance seeks to provide financial assistance by the term-lending financial institutions on concessional terms to entrepreneurs in designated industrially backward districts. The concessional terms include : lower rate of interest, longer repayment period and moratorium on loans, lower service charges and income tax rebate. The scheme is applicable to 246 districts, nearly two-thirds of the country. These districts are also eligible for the benefits under the Differential Interest Rates Scheme that is, loans at low interest (4 per cent) to specified low income groups, thus also catering to the needs of the small artisans and craftsmen in the areas.

The second scheme is called the Central Scheme of Investment Subsidy, under which an outright subsidy amounting to 15 per cent of the fixed capital investment not exceeding Rs 1.5 million is granted to industries** started in designated backward districts. 101 districts are covered under this scheme.***

* The reference is to the Wanchoo Committee which recommended certain financial and fiscal incentives to remove industrial backwardness.

** These may be new units or existing units undertaking substantial expansion.

*** The scheme has undergone numerous changes since its inception in 1971, both in area coverage as well as in the quantum of subsidy provided. Thus from the original 44 districts, the coverage has now been extended to 101 districts. The subsidy has also been increased from 10 per cent of the fixed capital investment to 15 per cent. Other conditions governing the grant of the subsidy were also relaxed.

Besides the Investment Subsidy Scheme, a subsidy is also given by the centre equivalent to 50 per cent of the transport costs of industrial raw materials and finished products between the location of an industrial unit (new or existing units undertaking substantial expansion) in selected backward and remote areas and the nearest railhead. This is a measure of inaccessibility neutralisation applicable to the states of Jammu and Kashmir, the North-eastern Region (all states and Union territories), Himachal Pradesh, the Hill districts of Uttar Pradesh, the Andaman and Nicobar islands and Lakshadweep.

Besides the various central schemes mentioned above, most of the State Governments are also offering certain incentives for promotion of industries in their backward areas. These incentives relate mainly to sales tax, octroi duty, power and water rates etc. A few State Governments have also taken up schemes for providing investment subsidy in selected backward districts and have set up corporations for development of infrastructure facilities. In 1972, a committee[36] made calculations to get a broad idea of the overall element of incentives available to an enterprise in the backward area by combining the incentives offered by the Centre and the States and found that this varied between 57 per cent and 66 per cent of capital investment i.e. between half and two-thirds of the investment.

The main object of these incentive programmes is to confer a certain competitive advantage on the backward areas so that private capital can be attracted to them. But in their actual operation, a number of weaknesses have appeared which have cast doubts as to the utility and efficacy of such programmes in securing a regional dispersal of economic activity and thus closing the regional gap.

An important point about the incentive programmes is their spatial coverage and the way in which the designated areas for the incentives have expanded with the years, rendering them *almost* neutral with reference to location, serves to highlight the difficulties in operating such programmes in democratic countries. For instance, the investment subsidy scheme was originally limited to 44 districts. Later, it was extended to 101 districts. Thus by spreading the resources thin over a large area, the necessary intensity of effort is not achieved. From other cosiderations too, the incentive structure, as it operates, does not seem to be quite conducive or designed to attain the objective of regional closure. We may list below some of the other weaknesses of the incentive schemes :

(i) An incentive must be sufficiently discriminating as to the constraints/problems which it seeks to pacify or neutralize. In one area, it may be offsetting the higher transport costs; in another, it may be offsetting the lower productivity of labour skills; yet in a third, it may be neutralizing the lack of entrepreneurial or managerial abilities. Thus the differential needs of the areas must be interwoven into the pattern of incentives.

(ii) The schemes are not sufficiently selective about the types of industries to be promoted in specific areas. This selectivity about the type of industries to be promoted is an important aspect for consideration, because in backward areas it would be necessary to create employment in occupations related to availability of natural resources and local skills.

(iii) A characteristic of the incentives in democratic countries is that once given, they tend to perpetuate themselves. Each incentive is evolved in the context of particular circumstances and therefore over the years, it has to be modified, replaced or withdrawn in the light of experience, changes in circumstances or intentions. But this becomes difficult, as vested interests grow and political pressures build up against any modification, replacement or withdrawal of the original scheme. This seems to be an inherent weakness of all incentive schemes.

(iv) When incentives are given to areas in different levels of development, they also tend to have a differential impact, with the more developed among them reaping greater advantages from them, thus further outdistancing the other areas. In this way, imbalances only grow rather than lessen. The Indian experience clearly shows that bulk of the concessional assistance has been shared by a limited number of districts in the more industrialized states like Maharashtra,

Gujarat, Tamil Nadu and Karnataka, thus accentuating inter-state imbalances.

In view of the above weaknesses experienced during the operation of incentive schemes, it would appear that for developing countries like India, it would be more effective if the measures for regional dispersal of industries are conceived in the form of more *comprehensive assistance*, including a technical and management input, *than merely as fiscal and monetary concessions*. This viewpoint has been appreciated in India and indeed some efforts on these lines have also been made. It will be therefore useful at this juncture to elaborate this experience. A case study of one of the Area Development Corporations set up in Maharashtra—The Marathwada Corporation—affords interesting insights in this context. Venkatesan has documented this experience (Venkatesan 1978).[37]

Marathwada is a relatively backward region in the State of Maharashtra. The Government of Maharashtra set up in 1967 a development corporation called the Marathwada Development Corporation in response to a demand from that region for a separate agency which would engage itself exclusively in the development of the region. This corporation adopted a strategy of appropriate management for ensuring a viable production programme to as many rural persons as possible by providing them with access to finance, means of production, technical and managerial know-how and marketing. This was attempted through a managerial set-up to ensure the access to the development package mentioned above. The various steps in the strategy are :

(i) Identification of the target groups in the area, the target group being defined as a group of people who are not only poor, but also relatively homogeneous with respect to the effects that a given set of policy instruments might have upon them, for example, small farmers, village artisans etc;

(ii) Identifying the type of economic activity which can be selected for each target group;

(iii) Detailed project planning; and

(iv) Providing the target groups with access to finance, means of production, technical know-how and marketing.

Depending on the project planned and the capabilities of the target-groups, suitable provisions are made for the upgrading of their skills and upgrading of their production techniques, thus introducing appropriate technology. The primary objective is to create more efficient producers in the rural sector, rather than wage-earners or in other words to solve the access problem of the rural self-employed sector. The appropriate management, strategy considers access to five essentials to uplift the actual as well as potential small producers in any area namely, credit, inputs and raw materials, technical assistance, managerial assistance and marketing of output. The strategy also attaches great importance to *the principle of coordinated provision of these services at the door step of the small man.* This is essentially an area management approach which needs careful planning and implementation as well as the careful surveillance (that is follow up) of the plan.

The experience of the Marathwada Development Corporation shows that the above ideas can be increasingly adopted in several rural development projects.

In India, an approach somewhat close to this strategy is being introduced on a massive scale in the rural areas. This is called the District Industries Centre Programme. The District Industries Centre seeks to provide an institutional framework that can bring a total integrated approach to the development of the industrial resources in the district, both human and material. Although it contains several elements of the "appropriate management approach", it is only an extension of the administrative framework into the area of industrial administration. As the experiment is still new, it is difficult to comment on its suitability and adaptability to the needs of the backward areas.

Growth centre policies

These essentially involve a combination of the characteristics of modern industries and the creation of agglomeration economies in an attempt to reduce the backwash effects between the backward area and the more developed regions and at the same time create spread effects to the growth centres' hinterlands. The implantation of modern industries in the resource rich but underdeveloped areas in India provides an excellent illustration of the effect of these policies. We have noted earlier in our discussion that such modern industries have only tended to create geographical dualism, and increased backwash effects and to disrupt the social life of tribal communities living in the rural areas. Therefore in resource-rich tribal regions, where fast industrial development is being forced on national considerations, Sharma has advocated a policy of cautious incrementalism and comprehensive planning whose dimensions will encompass (i) a longer time perspective; (ii) a much larger geographical area than occupied by the core industrial activity; (iii) a broader spectrum of economic activity irrespective of the fact whether it formally belongs to the core, organised or unorganised sector; and (iv) inclusion within the ambit of planning all social, economic and legal institutional aspects of the tribal system directly affected by the new economic activity in the industrial core.

The above approach to growth centre planning in sensitive tribal area is being appreciated as a rational course of developing such areas. Thus while an aggressive industrial growth centre policy is being discouraged in certain areas of the country, the establishment of small rural growth centres to make optimum provision of economic and social services to people in a given spatial area is increasingly being advocated.[38] In the latter approach, the growth centre is not seen as a substitute for a strategy which tries to identify investment and policy priorities, but a framework for realising these priorities. Thus in India, the growth centre planning has been advocated as *one* element of a sub-plan or area plan.[39] The massive block planning introduced in the country includes this element as an essential ingredient to correct the spatial injustice in the provision of basic services to the country-side.

The basic needs-oriented approach

While the potentialities for growth may vary in the different backward areas, at least *equity* in the distribution of minimun social consumption is necessary to ensure a tolerable standard of human welfare. This concept has influenced the formulation of a national Minimum Needs Programme (MNP) in India in which the backward areas of the country have been given a specially favoured treatment. A minimum needs programme implies the identification of the front-line efforts, that is, the selection of the most essential requirements or needs for an area. The Fifth Plan (1974-79) set out an ambitious programme for the provision of drinking water, house sites for the homeless villagers, access road, elementary education, rural health, rural electrification, environmental improvement of slums and nutrition programme. But during the plan period, the outlay originally earmarked for this programme had to be reduced on account of the inflationary crisis of 1973-75. Systematic area planning which should have preceded this effort could not be made all over the country. Thus the implementation of the MNP programme did not come up to expectations during the Fifth Plan. During the Sixth Five Year Plan (1978-83), more than a five-fold increase in the outlay has been envisaged. The programme component includes all the basic items included in the Fifth Plan, besides adult education.

In a minimum needs programme, we are concerned about equity in the distribution of social consumption. This involves a systematic effort to locate the minimum needs at appropriate points, where they can be used or act most effectively to achieve commonly accepted purpose or goals. Hence locational aspects and spatial planning become important issues. In such spatial planning, the population to be served and the distance at which services should be provided become crucial issues in decision-making. In order to achieve equity in a spatial sense, the norms should be area specific. India's programme of minimum needs however, seeks to impose certain national norms in this respect. The block level planning that has been advocated at the local area level is expected to take care of the locational aspects and to introduce some flexibility in the operational designing and implementation of the programme.

The target group approach

Yet another direction in which the specificity of Planning activities in an area has been sought to be increased is through its focus on selected target groups in the area. In this approach, the problems of the target groups in the community are intensively studied and solutions brought to bear on helping to over-come these with a view to enhancing their development opportunities and incomes. The small and marginal farmers programme is an example of this approach. In these programmes, the disadvantaged segments of the farming community in an area are identified and they are enabled to participate in the development process through specific programmes formulated for their benefit. An agency is created for the purpose* whose main task is to assist the small and marginal farmers to obtain credit and other requirements to develop their farms. They are to function primarily as catalysts and foster coordination between beneficiary participants, and various institutions, development departments and extension agencies. The programme elements include a number of activities like the promotion of adoption of HYV crops, provision of agricultural implements, facilitation of land development programmes, creation of minor irrigation works, provision of irrigation pumpsets, encouragement to schemes like sheep rearing, dairy enterprise and poultry farms and fostering of the growth of rural industries. At present, there are 169 projects in the country and the Government of India have invested about Rs. 113 crores on this scheme from its inception up to March 1977. Although the extent and intensity of coverage under this programme leaves much scope for improvement in a country like India, yet the various evaluations have accepted the favourable impact on income and employment situation in project areas. In the working of the programme, however, a number of deficiencies and limitations have been noted.[40] Some of these are inherent in the coverage and scope of the programme; others pertain to its working.

The SFDA programme has, in a large number of cases, remained a scheme of institutional credit. It is now sufficiently recognised that unless these credit progrommes are properly

* For these programmes, the Small Farmers' Development Agencies have been set up. Each agency roughly caters to one district and is a registered Society. The main objective of the agency is to enable the small, medium farmers and landless labourers to participate in the developmental process "to prepare appropriate programmes, help ensure availability of inputs, services and credit, and evaluate, the progress from time to time."

linked up with transmission of technology, input supply, marketing facilities and necessary infrastructure, the progress acheived by the target group will be slow; in some cases, it may even be injurious, as the farmers are likely to be caught in debts without a repaying capacity. As conceived at present, the Indian programme, is in the nature of an isolated programme insufficiently integrated with other development programmes in the area and not supported by any meaningful area planning. Whereas in the DPAP the *area* has a resource-use dimension, for the SFDA programme, the area is merely a convenient unit for the identification of small farmers. Unless the relationships with the other local resource-based programmes and activities in the area are consciously forged and a basis for self-sustaining increases in productivity and income is ensured, the programme will tend to remain, by and large an ad-hoc measure without contributing to sustained growth. Thus an important lesson learnt during the implementation of this programme is that area planning should be the vehicle through which beneficiary-oriented programmes of this nature should be operated. An approach based on such a concept is being implemented for the tribal areas in India.

Resource problem-oriented approach

Structural handicaps or inefficient use of infrastructures may tend to result in areas remaining underdeveloped, although national resources may be present. The drought prone areas is an example of the former t,pe, while areas coming under the command of large irrigation projects constitute an example of the latter case. Nearly 19 per cent of the total area of the country with about 12 per cent of the population (consisting of 54 whole districts and contiguous areas of another 18 districts that is 72 districts either partly or wholly) belong to the category of Drought Prone Areas. Till the middle of the Fourth Plan, the problem of such areas was largely tackled through the device of scarcity relief works, which were in the nature of crash programmes, akin to fire-fighting, which were resorted to as an emergency measure, as and when scarcity conditions appeared in any area. Such programmes were confined to limited activities viz. minor and medium irrigation, afforestation and soil conservation, roads and drinking water supply, which failed to effectively link up with the economic activities in the area. The idea of tackling the problem on ecological lines, adopting a resource-based and area specific approach and integrating various elements for uplifting the economy of the area as a whole, represents a major departure in the concept of development of backward areas. This change in concept resulted from the report by a Task Force[41] which went into the various aspects of integrated agricultural development in drought prone areas. This Task Force recognized that the various drought prone areas were not homogeneous land and water resources situation[42] and that a uniform strategy of development cannot be made to fit all of them. The ingredients of the programme must be increasingly area specific, taking each watershed into account and working out the programmes in accordance with the land and water resources situation. The approach advocated was that the main thrust of effort must be directed towards the restoration of a proper ecological balance in these areas with the following elements constituting the strategy viz. development and management of irrigation resources, restructuring of cropping pattern for pasture development, changes in agronomic practices, livestock development, provision of drinking water supply development of rural communications and development of small marginal farmers and agriculture labour.

The DPAP (Drought Prone Areas Programme) marks a definite departure in our conceptualisation on backward area development and is oriented to bring the programmes of development closer to the problems of the area. The strategy has led the planners to take a close look at various aspects of agricultural development in relation to the area, aspects like intercropping, multi-cropping, agro-climatic zones, water-shed management, farm forestry, fodder and grass land cultivation, scrub jungle cultivation and dry land farming. Suitable programme have been devised for the proper management of scarce land and water resources, taking into account the problems in shifting from low value to high value crops and to conversion of poor land from farming to pasture. During the operation of the programme, the need for more effective community and group action for the effective transfer of technology, development of local resources, certain common activities like community nurseries, as well as for the development and management of certain common aspects has been realised. Being a central programme, the DPAP has tended to operate in a somewhat isolated or exclusive manner from the general programmes in the area. The need for more effective integration of the DPAP with other economic programmes in the area and the basic needs oriented programme for the community has been increasingly realised. Steps are being taken now to integrate them into the overall Block level planning for full employment programme.*

Command area development

One of the interesting aspects of the developmental experience in India is the fact that some of the potentials created by the development projects remain un or under-utilised, perpetuating the condition of backwardness in the hinterlands of those project areas. This was the case with a large number of major irrigation projects whose potential remained under-utilised largely because some related works like construction of field channels or infrastructure like roads and so on were not simultaneously planned and executed. In order to realise the full potentials of these projects, a Command Area Development Programme (CADP) was undertaken during the Fifth Plan. This is basically a coordinated development approach involving various related operations like the construction of field channels, land development, consolidation of land holdings, carrying out of soil survey, determination of cropping pattern, supply of inputs, research and development of scientific farming to help progressive agriculture and provision of various infrastructure such as roads, railways, markets, warehouses, electricity and other facilities. There are now (as on 31st March, 1978), 38 Command Area Development Authorities covering about 14 million hectares. While no assessment in quantitative terms of the benefits accruing from the efforts made so far has been made, the yield figures available for some command areas indicate that crop yields have increased from 1½ to 3 times in these areas.[43] Thus the Command Area Development projects have been able to achieve much. One of the weaknesses noticed

* The Block-level Planning programme has been discussed under (x) Integrated Rural Development in this Chapter.

about them, however, is the lack of integration with the general programmes of the area particularly with many other related activities like the development of processing industries, provision of community services etc. Such type of integration is now sought to be achieved by dovetailing this programme into the country-wide block-level planning programme for full employment.

Asset redistribution policies

Among the policies aimed at direct redistribution of existing assets, those relating to land claim first attention, as land is the major means of production in the Indian economy. It is but fair to mention that the Zamindari abolition has been totally accomplished, which has conferred on the farmers distinct benefits in terms of their rights to land and enjoyment of the fruits of their labour. Land reform legislations have been enacted in all the States, but they have left many loop-holes with the result that it has been difficult to implement them effectively. The land located as surplus for redistribution is very low (far below expectations) and land redistribution to the poor much less. An overall review by Vyas[44] shows that as a result of the legal, demographic and market forces, the hegemony of large landholders is challenged, and ownership has become less skewed.

Among the indirect measures of redistribution of income, the substantial expansion of public ownership over the last more than two decades must be counted as an achievement. The public sector now owns more than 25 per cent of the total reproducible tangible wealth in the country. According to Bardhan,[45] it is doubtful if this has had any significant impact on personal income inequalities.

The approach to hill area development

The sub-plan approach

The hill areas constitute another type of lagging areas in the country where a multi-directional approach to area development has been accepted as the policy for accelerating the development of these areas. Among the hill areas, some constitute separate States like Himachal Pradesh and Meghalaya. Some others constitute parts of States which are hilly where the developmental trends have tended to favour the plain areas at the cost of the hill areas. In order to assist the development of this latter type of areas lying in Uttar Pradesh, West Bengal (Darjeeling Hill Areas), Assam (Mikir & N. Cachar hills districts) and Tamil Nadu (Nilgiris), a sub-plan approach* has been followed since the Fifth Plan. This ensures that a certain order of development effort flows to these backward areas from the State funds which is supplemented by special additional central assistance. For this total investment available, a plan of development is drawn up for the area in which the resource development programmes are adequately balanced by essential infrastructural provision, particularly communications and the minimum needs programmes like the provision of drinking water. The main objective of the hill areas sub-plan is to orient the programmes specifically to the local conditions so as to benefit directly the people living in the hills. Priority is being given to the schemes which are employment-oriented and are based on the maximum utilisation of local resources. The guidelines issued by the Planning Commission have emphasised that the nature of the linkages and interactions between the plains and the hill areas should be identified and reflected in the sub-Plan. In particular the following types of linkages have been mentioned :

(1) Development of sheep, cattle, poultry and dairying with marketing, pasture development and provision of health cover;

(2) crop husbandry and plantation crops with soil and moisture conservation and land management practices; and

* The sub-plan approach in the Indian context confers three definite advantages : (1) it is likely that because of their weak organisation and bargaining strength, the target groups in these areas did not get their proper share in State development plans. The sub-plan pointedly draws the State's attention to the share given to the areas in question in divisible plan expenditure, (ii) the sub-plan enables the centre to have a direct voice and interest in the areas in question because of the financial assistance given by way of an additive, (iii) much more than the points mentioned above, a sub-plan within a State plan means that the special problems of the areas receive specific attention.

(3) development of horticulture and forestry with programmes relating to the opening of inaccessible areas and marketing.

The sub-plan approach to hill area development has encouraged the development of plantations (for example, cinchona), horticulture, sericulture and animal husbandry in specific areas suitable for such development. Certain specific problems of the hill areas like shifting cultivation, soil erosion, minor irrigation etc. have also been tackled. But the hill area problem has not yet been sought to be tackled on a long-term basis. The sub-plans are essentially medium-term plans. Some techno-economic surveys have been conducted to reveal the resource potentials. But no systematic attempt to prepare perspective plans for these areas and derive the medium-term and annual plans from such perspective has been made so far.

The approach to tribal underdevelopment

The areas of tribal concentration constitute the most backward among the backward areas in India. The Government of India's approach and strategy to tribal underdevelopment represents a bold attempt to tackle this problem by ensuring adequate weightage to the local priorities in the State level programmes. It combines both functional programmes undertaken to remove a single constraint that is considered particularly critical in the context and area programmes involving a multi-facetted approach to rural development.

A two level planning activity has been envisaged in this context. The Tribal Sub-Plan is the first level, which is conceived as an integrated area development plan for designated areas of tribal concentration, representing total developmental effort in these areas. The second level of planning is at a further micro level, known as Integrated Tribal Development Projects (ITDPs). The entire sub-plan area in the country has been divided into 180 ITDPs. Thus the ITDPs provide a frame for family-based micro-plans. The sub-plans together with the ITDPs represent a comprehensive approach to influence the developmental process among the poorly integrated societies. The tribal areas, particularly in central India form part of bigger resource regions—regions of unexploited resource potential and are located on the borders of different states. It has been realised that the sub-plans and the ITDPs, by themselves, are not adequate for determining the direction and pace of development of the bigger regions, particularly because of big-sized mineral and industrial development investment decisions involved. A Working Group on Tribal Development for the Sixth Plan has recommended that a broader planning framework at regional levels must be evolved so that there may be full harmony between the efforts at the micro and macro levels.

The sub-plan strategy lays high stress on administrative structure and personnel which are fundamental to effective implementation. Several adaptations which are necessary in this context to make the programmes conform to the specific problems of these areas have not really come about. There is a considerable degree of lag in the delegation of administrative and financial powers to the Project authorities in keeping with the responsibilities assigned to them. The inter-departmental coordination which is important for achieving the integration of sectoral programmes at the ITDP level, however, has not come obout in many areas. The level of investments in the tribal areas has no doubt gone up. The experience during the Fifth Plan has, however, brought to the fore certain constraints in relation to the efficacy of the delivery system as also the absorption capacity of the tribal communities themselves. A close review reveals that the bigger programmes and infrastructural development have stolen a march over the small programmes of direct relevance to the tribals. Obviously administrative preparedness appears to have come up as an important constraint in the effective implementation of tribal development programmes and protective policies.

The backward area problem in the tribal areas faces the biggest challenge of financial, manpower and institutional constraints. It is only fair to add in this context that few countries in the world have made more efforts than India to protect the rights of their indigenous peoples and to integrate them into national development planning. Notwithstanding the fact that a big break-through has been effected, the progress of adaptation of programmes and their implementation in different sectors is uneven and serious gaps are noticed between planning and actual project performance. Critics have commented on the various aspects of underdevelopment in the tribal areas, particularly on the social costs of resource development in the tribal areas and have advised to "hasten slowly". Perhaps the theory of Development continum, implying a sequential approach, has much relevance in the tribal areas than in any other type of situation.* This particular context calls for both structural as well as programmatic remedies. A conscientisation process has to go hand in hand with a broad-based development of the subsistence sector and building up of technical and institutional capabilities. A few simple interventions have undoubtedly been made to remove the most critical constraints and the programme must now evolve in scope through time-phasing of activities.

Regional planning approach for special problem areas

In some critical areas where some special problems have arisen, Government of India has resorted to some kind of a regional planning approach to solve such problems. The north east region of the country and Andhra Pradesh afford instances of this kind.

The north-eastern region is in the nature of a peripheral area, isolated from the major centres of economic activity in the mainland. Most of the States in this region are of recent origin and economically not viable. They have inadequate development administration and technical competence for undertaking development as compared to the situation obtaining in other parts of the country. Since a big gap is to be covered as quickly as possible in this context, the centre has provided for a regional plan as an additive component to supplement the normal plans of the States. This regional plan is mainly an infrastructural plan consisting of important inter-state projects such as power, flood control or transport and communication projects, regional educational and training institution, regional marketing institutions and so on. A north-eastern council set up under an Act of Parliament in 1971 coordinates the regional planning effort. The scheme has been

* Attention is drawn in this context to Sharma's model of development of small tribal community discussed in Chapter 4 of this book.

under way during the Fifth Plan and it is too early to quantify the extent of regional closure achieved by this method.

Another case of special intervention through regional planning approach to tackle the problem of regional imbalances is that of Andhra Pradesh. This State consists of three major regions, among which regional disparities were striking; which led to serious political unrest in 1973. With a view to maintaining the political integrity of the state and to bridge the yawning regional gap, the Government of India made available a special central assistance of Rs. 90 crores for the accelerated development of the backward areas in the State during the Fifth Plan period (1974-78). The funds were to be allocated to the three regions (Telangana, Rayalaseema and Coastal Andhra) in the ratio of 5 : 3 : 2. Planning and Development Committees have been constituted for each of the regions and these committees have been preparing the plan and implementing it. The backward areas were determined on the basis of identification studies using relevant indicators.[46] In the absence of detailed evaluation studies, it is difficult to say to what extent the approach has led to reduction in regional disparities.

Integrated rural development for full employment through block level planning

This represents a small area based approach. The heart of the backward area problem is that solutions must be found essentially in their local settings, taking note of the local peculiarities, potentialities and needs of particular areas. This would mean that planning should become increasingly area specific. Such area planning would enable the selection of programmes suitable to the particular area, and more importantly, the identification of available and potential labour, the training and skills required for them and tying up such identified labour with labour absorbing programmes. Area planning would also enable the effective integration of different programmes in the area and will secure the optimisation of benefits from each programme. In this manner, area planning will seek to establish a greater degree of consonance between the problems of the area and the development programmes suited to their milieu.

In India, there has been a continuing effort to build plans from below, right from the Second Five Year Plan onwards; but various problems including problems in planning methodology, constraints in technical personnel resources, the inadequate degree of decentralisation of planning and decision making functions to the local levels and the overall constraints of financial resources have inhibited the local planning efforts. In regard to micro-level planning methodology, significant contributions have come from some pilot projects. Among them, the methodology for service centre planning, credit planning and employment planning may be specially mentioned. During the course of several experiments in local planning, it has also become clear that questions of social and political structures and processes need to be considered *pari passu* with economic policy and methodological refinements in planning. Hence great emphasis is now placed on laying down a social and institutional basis for development and in making decentralisation of planning and decision-making functions effective, so that local initiatives will determine the development effort. However, the required degree of decentralisation of political, administrative, technical and financial powers to the local levels has not yet come about. In the matter of allocation of resources, both at the national and state levels, increasing weightage is being given to the degree of backwardness of an area. For the allocation of resources to the districts from the State, some State Governments have devised suitable criteria, which takes into account the degree of backwardness. Thus, increasing weightage is being given to backwardness in the matter of allocations of resources to the district levels.

The policy shift which took place recently has given greater importance to grass roots level planning and bottom-up strategies. Thus, more and more area specific planning, which will be local resource based, basic needs oriented, labour intensive and employment oriented and based on appropriate rather than highest technology is being advocated to remove the poverty in the rural areas. The development block, which has been identified as the unit area for planning, becomes the areal frame within which all on-going programmes in the area will be adapted, integrated and unified. Thus, the present block level planning approach may be seen as a phase of consolidation and integration of the several differentiated programmes which were evolved during the previous phases.

In the operational guidelines for block level planning prepared by the Planning Commission,[47] resource inventory of the block, project identification and planning, manpower planning and budgetting, household planning for families below the poverty line and spatial planning for minimum needs and socio-economic infrastructure stand out as the major components in the Block-level planning programme. The household planning for families below the poverty-line which is an important component of block-level planning follows the Antyodaya approach, which is defined as upliftment of the last man. Some experiences exist in India in the application of the micro-level planning technique for planning the poorest of the poor. We may draw particular attention to three important experiments carried out in different parts of the country.

One of them is the case relating to the preparation of an Integrated Area Development Plan for six blocks falling in Ghazipur-Ballia districts in the backward belt of Eastern Uttar Pradesh (Anand Sarup 1978).[48] The project arose out of an emergent socio-political event in which the lower castes (called Harijans) clashed with the landed aristocracy (the Bhumihars) which led to deep social tensions, leading to the former being thrown out of the traditional agricultural occupation as agricultural labourers. This provided the motive for change to allied occupations. The target group for the project includes all persons with a land-holding of less than 0.5 hectare or families with incomes of less than Rs. 2400 per annum. The project was started in 1976. The major activities incorporated into the project are : the distribution of milch cattle and improvement of their breed; sheep, goat and pig-rearing; poultry-keeping; pisciculture; planting of early maturing fruit trees; cultivation of high value spices and vegetables; distribution of agricultural inputs and improved implements; provision of irrigation facilities jointly by farmers owning less than 0.5 hectare; development of cottage industries like tanning, clay pottery, rope-making; establishment of production-cum-training centres for blanket-weaving; and sewing and embroidery for rural women. An up-grading of the level of social consumption was envisaged through provision of drinking water, establishment of dispensaries and maternity centres, training of mid-wives and encouragement of house construction by the poor. Programmes for infrastructural development included extension of communications, improvement of drainage and construction of small-scale flood protection works.

For the implementation of this pilot project, a very flexible administrative pattern has been adopted. An authority with representatives of the beneficiaries, local institutions and the administrative and technical bureaucracy has been created which can appoint suitable personnel, sanction expenditure, reallocate original outlays, start new programmes, revise the plan from time to time and coordinate the functioning of all departmental officials functioning within its jurisdiction without a prior reference to any higher level of government. This, as an administrative innovation, is quite unique in the Indian system.

With only three years that have gone by since the project was launched, it is difficult to comment on its performance. But from an Interim Evaluation Study conducted in November 1977 and based on periodical evaluation reports available, the following tentative conclusions may be stated :

(i) It is extremely difficult to protect the interest of the weakest sections of society, because they are not organised and their voice is drowned by the more strident demands and assertion of better placed citizens, who appear to dominate all councils and committees;

(ii) Formulating short gestation programmes for diversified productive employment on a widespread scale within the constraints of limited expertise, small resources, lack of entrepreneurship, poor mobility and a skewed and thinly spread infrastructure is no easier than planning for large-scale enterprises;

(iii) Without widespread delegation of powers, local agencies charged with planning and implementation can usually get nothing more than lip-sympathy from departmental functionaries;

(iv) Nothing is likely to be achieved in ameliorating the conditions of the poorest among the poor without a great deal of procedural flexibility and risk-taking;

(v) Very dedicated work on the part of the Government functionaries is called for and a radical change in their motivation and orientation right down the line is necessary;

(vi) In the ultimate analysis, development of backward areas as well as disadvantaged groups will involve a net inflow of substantial resources till they reach a minimum threshold of development.

The second case of Integrated Rural Development Project with an explicit target group approach in which detailed household planning was attempted to elevate the people below the poverty line is that of Somangalam in Tamil Nadu. The Project Area is a cluster of eleven villages around Somangalam (central village) which is in close proximity to Madras city. The village cluster has 2238 households (a little more than 11,000 people) of which 225 are big farmer households and the rest belong to the weaker sections of society (small and marginal farmers, landless labourers and service households). In 1976 when the project commenced, the pattern of income distribution among the households was as follow :

(i)	Less than Rs. 200 p.m.	1398 households
(ii)	Rs. 201-400 p.m.	606 households
(iii)	More than Rs. 400 p.m.	234 households.

It was found that in less than three years, the position was reversed. This miracle was possible through intensive household planning attempted by the Rural Development Department of the Government of Madras. The method was to prepare and implement for each household a plan for a period of three years in consultation with the head of the family taking into account the traditional occupation of the family, their felt needs, assets possessed by them, manpower available in the family and the skill possessed by them. The plan was tied to rural credit and the Development Staff assisted the households to obtain the credit, supervised each household project and ensured repayment to the bank as the project began to yield returns. The success of the project was due to this day to day supervision and the ability to marshal and bring together various forms of assistance to each household, virtually to its door step. Simultaneously with the household plan, other common aspects benefiting the entire community were looked into. Deepening and strengthening of the bunds of irrigation tanks, repairing link roads, preventive medical care for animals and human beings, post and telegraph facilities were some of the common items included in the village plan. The main thrust of the household plan was in animal husbandry programmes (particularly distribution of milch animals), agriculture (mainly provision of crop loans, assistance towards sinking of wells equipped with pump sets and provision of work bullocks in pairs) and industries programmes (mainly assistance to weavers). The Somangalam experience shows that through intensive household planning and effort to mobilise all government assistance to the target groups combined with adequate supervision, it is possible to raise the level of living of the poor people. But it requires very careful coordination at different levels, including the extension, management, production, credit and marketing services. The active support of all institutional agencies is very necessary for the success of the project.

A third experience to which we would like to draw attention is the case of the Antyodaya programme of the Government of Rajasthan, a State in the north-western part of India. The programme envisages in the first stage, the upliftment of 5 poorest families from amongst those below the poverty line in each of the 33,000 villages of the State. To formulate the detailed methodology of the programme, a pilot study was launched in October 1977 in five selected villages in each of the districts of Jaipur, Jhunjhunu, Chittorgarh, Kota and Jodhpur. In the first instance, the families below the poverty line were identified based on the following criteria :

(i) Families under severe destitution having no economic assets and no member in the age group of 15-19 years capable of economic activity;
(ii) Families having no economic assets but having one or more persons capable of economic activity where annual family earnings do not exceed Rs. 1200;
(iii) Families with some assets whose annual income is below Rs. 1800/- per annum; and
(iv) Remaining families having some land and assets, but

are below the poverty line that is below per capita consumption level of Rs. 55 per month.

The beneficiary families for the first stage of the programme (that is the bottom 5 families in each village) were then identified by the *Gram Sabhas* (village councils). The needs of these families and the possible ways of assisting them to achieve a reasonable income level were then discussed taking into account the social and economic profile of the families and also of the individual preferences of its members. A plan of action[49] was then drawn up by the Block Development Officer assisted by the extension staff and then consolidated to form the District Antyodaya plan. The active involvement of the financial institutions, including the commercial banks was secured and the programme content of the Action Plan for individual households was tied to the assistance provided by them. Although the programme was launched in October 1977, it gained momentum only after July 1978. It is, therefore, too early to assess the benefits. Nevertheless, a quick evaluation conducted by the Planning Commission[50] revealed that apart from providing gainful employment or an economic activity to the beneficiaries of this programme, it had also helped in raising their status and credit-worthiness in the village. The Rajasthan Government look forward to creating a wide range of employment opportunities through this programme by which some means of livelihood will become available to the poorest of the poor.

The three case studies referred to above were brought into this discussion mainly to highlight the potentials of a small area planning approach for Integrated Rural Development in the developing countries, into which is meshed a consciously worked out target-group-oriented approach as an important and inseparable component. If an Area Plan is interpreted only in a limited sense of an infrastructural plan, then the controversy between place prosperity and people's prosperity[51] will hold good, as such a plan will tend to benefit only a section of the population, marginalising the weaker sections and thus accentuating social imbalances. Our contention is therefore that an Area Plan for a developing country should also include household planning for the target groups as an essential and inseparable element. In this way, the spatial as well as inter-personal inequalities can be simultaneously dealt with and growth with redistribution realistically achieved.

India's Block level planning programme is conscious of the above truth and it therefore proposes to select in each block 1500 poorest families and help them in setting up viable economic activities including rural industries and services and thus enable them to cross the threshold of poverty line. These 1500 families will be selected in a phased manner over a period of five years. It is estimated that a subsidy of Rs 1500 per person along with the bank loan that it will enable him to get, will be sufficient to create a viable work opportunity for him.[52] Simultaneously, the block plan would have worked out the production and infrastructural elements of the plan into which the household plan for full employment will be dovetailed. If this programme gets going, then a breakthrough towards an autonomous process of narrowing down of spatial and inter-personal disparities can be made.

To sum up, India's current approach to backward area development is rooted in an Integrated Rural Development approach. Such Integrated Rural Development is based on an area development approach in which household planning is a distinct component. Since the intended transformation cannot be achieved overnight, it is sought to be achieved in stages, wherein the development of the local resources through local initiatives (that is local participation) and the elevation of the poorest of the poor above the poverty line are consciously aimed at. A self-reliant and self-sustaining development is the ultimate goal. The approach towards this goal seeks to combine many strategies simultaneously—that is growth-oriented, employment-oriented and poverty-oriented strategies. The appropriate mix of these strategies and the programmatic details will depend on the particular area situation; but if worked out satisfactorily, can help to reach the ultimate goal of growth with equity with least friction.

Lessons from the Indian experience

We may now pull together the threads of this narration on the various programme approaches to backward area development with a view to extracting some valid generalisations. The Indian approach to backward area development appears to be characterised by three important elements, namely selectivity, specificity or focus and dynamism or flexibility. An all-at-once approach to the problem has been discarded; indeed such an approach is not feasible. The approach, initially has to be one of making a deep thrust into a specific problem, thereby contributing to the first breaking-down of what has been identified as the key obstacle or obstacles to development in the particular area. Thereafter, the development strategy must add to this, other elements thus gradually enlarging the scope and contributing to greater and greater comprehensiveness in planning and development. This implies progression from the particular to the universal, from part to the whole. But such progression will not be along a straight line, so that no simple rules of progression are possible.

The Indian experience shows that the closure of the regional gap will be effective in a highly decentralised approach through micro level planning in which it would be possible to go area by area, taking advantage of local participation, local supervision, local insights and additional resource mobilisation. Attention to the target groups—the people below the poverty line—must be part and parcel of such planning; otherwise the weaker sections will tend to get by-passed. In large countries like India, planning has not only to be at the micro-level, but also at multi-level so that proper "nesting" of plans takes place and distortions do not develop.

The Indian experience also shows that the removal of backwardness is not necessarily achieved through lumpy investments alone; very often, by making several half-way arrangements and filling up gaps which may involve only small complementary investments, it would be possible to pep up the development effort in a backward area. Increasing the utilisation of existing capacities, already created in the area, is one method; yet another would be through deliberate efforts to capture the various inter-linkages and externalities in development. Improving management for better performance for example, Appropriate Management approach, is still another technique. Such interventions involving limited investments are likely to lead to visible improvements. Again, certain non-economic measures, like inducing basic social and institutional changes in the area, are very important for the development of backward areas. In fact, providing the appropriate social and institutional basis for development is regarded by some as the most important step in the backward area planning strategy.

Thus from the experience available in India, one may say that the method of planning for a backward area would be : first to define the local area for planning and then to identify an appropriate package of development programmes for the area. This would involve defining and locating the areas where (a) concentrated investments will be required, (b) utilisation capacities will have to be maximised, (c) certain gaps have to be filled by capturing economic interlinkages and externalities, and (d) certain basic social and institutional changes are required.

Having defined these parameters, it would be necessary to resort to specific household planning of families below the poverty line—the target groups—so that concerted effort is made to bring them above the poverty threshold. Thus the backward area development strategy appears to be one of adopting an area specific approach combined with a functional approach to reach specific problems and target groups. As set out in the beginning of this chapter, it is an area specific, people specific and problem specific planning and calls for a lot of coordinated effort.

The critical components of a development strategy for backward areas

A striking feature about the Indian experience on backward area development is its eclectic approach to the problem. The attempts cover a wide ranging selection of special programmes which are either resource problem based, or target group oriented. At times, they are limited to particular sectors as in the industrial incentives programmes; or they may be quite comprehensive in nature, as in the area development programmes evisaged for the hill and tribal areas. The emphasis on strategies is also different in the different situations—sometimes it is growth oriented strategies, sometimes family-based and poverty focussed strategies or basic-needs oriented strategies. More recently, the employment-oriented strategies have come to the forefront. The Indian experience thus suggests two propositions: (1) there can be no single and invincible solution to the backward area problem. Each area situation has to be analysed in its particular context and appropriate strategies must be devised. (2) The choice of interventions should be constantly reviewed and changed, as necessary, from time to time. Thus constant adaptations are necessary.

In the field of development planning, as in many other constructive fields, two different stages repeatedly appear; one, in which the development experiment makes some new dents, and the next phase when the knowledge and experience gained from such experiments are consolidated and systematically ordered. The Government of India's approach to the development of backward areas originated as several isolated and ad hoc approaches to the problem and has been gradually adding to this, various elements, calculated to bring in greater "comprehensiveness" in the approaches conceived, greater depth in the programmes devised and better "integration" among the programmes undertaken for the backward areas. The selective interventions originally devised, while making some dents into the problem, could not tackle it in all their multifarious aspects. Further, the special programmes tended to remain aloof from the other programmes in the area. The barriers to development in the backward area, in fact, proved to be too tough and intricate for the required degree of integration to be achieved autonomously; hence a deliberate integrated approach appeared to be necessary. The present trend towards block level planning, in which the on-going programmes are sought to be integrated and adapted to the new strategy, may be seen as a move in this direction.

Thus through a continuous process of trial and error, involving adjustments and modifications in the light of changing perceptions of the problem, we seem to have moved towards the next planning phase in search of greater comprehensiveness and better integration. If one sees this development drama with more than casual perception, a certain *sequence* in the development process is striking. It is apparent that the transformation of a backward area in developing countries, like India, cannot be effected overnight by any magic or developmental acrobatics. It has to be consciously striven *through time*, with some definite progression in a march through a development continuum. In other words, a transition strategy to backward area development seems to be vitally necessary, if the problem is to be systematically tackled.

The planning for backward areas is therefore, really a problem of planning for transition growth. Here we seem to have no choice but to plan, as best as we can, for a situation of continuous rapid transition. This implies a time element in the process. During transition, the backward area economy moves through radical changes resembling a process of evolution. This feature of transition growth suggests the need for identifying different transitional phases. The sequence of phases means that the economy is not only growing in the quantitative sense, but also that its basic rules of growth are being gradually modified. Also in the different types of backward areas, the patterns of transition will differ. Corresponding variations will appear among the different kinds of backward areas in regard to the sequence of development phases, producing different patterns of alteration in the economy's mode of organisation. Thus it is clear that the backward area problem cannot be straight-jacketed into one set of planning rules. Each type of backward area has to be subjected to a different approach and has to go through a different sequence of development phases. Thus a *typological approach* to the problem is indicated. When a typological approach is adopted, it would help to obtain a clear perception of the problm from the point of view of the factors of causation of the diversity.[53] In this manner, it would be possible to make an analysis of the effect of these background factors on the transition growth process and such analysis would assist in framing a logical plan of development for the area.

This brings us to yet another aspect relevant to planning in developing countries, where the backward area problem is both complex and diverse. Even within the boundaries of a single country, the backward area characteristics vary widely in respect of natural environment, history, race, culture, density and growth rate of population, social organisation, resources, economic development and organisation and agricultural systems and methods. Virtually, the only characteristic that these areas will have in common is the poverty of the great majority of the people.*

The planning for transition growth in a backward area necessarily implies that such transition should be characterised by a certain 'continuity' in the philosophy and approach to the problem and a certain 'consistency' in the use of the policy instruments devised for the purpose. The transition, however, should be conceived as a time-bound phase. In the Indian context, it may be construed as a period ranging from 10-15 years. During this period, the crucial bottlenecks to progress and the barriers to development must be identified and these should be sought to be overcome through the various means elaborated in the plan. Thus planning for successful transition growth would require the economy to be carried through a succession of growth phases, in which progressively more and more advancement will be ensured. Each emerging phase of development will be conditioned by the growth accomplishment in the preceding phase. A realistic development strategy must therefore, ensure continuity in the policies, approaches and programmes and must ensure that the present is linked to the past and then to the future. No hard and fast generalisations are possible concerning the evolution of backward areas during the transition phase of development. For each type of backward area, it is possible to draw up certain ideal-type directions of structural change and incorporate them into the plan.[54] This chapter however currently does not propose to go into those aspects.

In the backward areas, the individual project planning also, must be governed by a policy of gradualism and cautious adaptations to change, due to the limited resources available for development on the one hand and the lack of information on the actual impact of development measures on the other. This approach sometimes, termed the Point-Line-Network-approach—requires that innovations are first tested for feasibility at one point and then modified, if necessary. In the next phase, this innovation is introduced along a line in different locations that is under different environmental conditions,

* As noted in the earlier part of this chapter, 'Underdeveloped India' is further divisible on problem lines into five major types.

in order to identify the necessary regional or local modifications. Once this phase has been completed, the innovation can spread in a network over the whole area as soon as possible.

From what is stated above, it is clear that the transition growth approach to backward area development also implies that a more forward looking perspective is always needed to consider the emergence of a new growth phase in the future. The Indian experience does not appear to have paid adequate attention to this aspect of planning for the backward areas. Of course, in the area planning approach, currently being emphasised in the country, one sees an attempt to systematically order or synthesise the different backward area strategies and programmes into one consistent set of policy instruments operating in a specific small area, that is the development block. But the strategy seems to be basically concerned more with the short run solutions to the problem, rather than with a long run view of transition growth. It is therefore, emphasised that in order to tackle the backward area problem more systematically, it is necessary to adopt a long term transition growth strategy. This will ensure that the short-term solutions fit into a long term strategy.

While emphasising the micro-level approaches to the planning of backward areas, one should not lose sight of the other spatial perspectives. It is well known in planning that the worm's eye view has to be matched with the bird's eye view. In a sub-continent like India, multi-level planning is needed and an important problem in this context is the nesting of plans. Thus the present package of bottom up strategies will have to be harmonised with the top-down strategies. In this manner, multi-directional approaches, multi-dimensional perspectives and multi-level planning are to be brought to bear on the problem of development of backward areas.

In the planning for backward areas, the location, intensity and direction of the initial trigger to the development process is of crucial importance. Such an initial trigger may be envisaged as occurring in some specific sectoral or physical space, which may have some *comparative* advantage for investment programming. This is an important consideration because a heavy pre-occupation with the backward area strategy in the earlier development stages in a developing country is supposed to have a dampening effect on the national growth rates. It is, therefore, necessary to have some distinct sectoral or spatial and locational preferences for investments, even within a backward area so as to take advantage of the comparative advantage theory. In this manner, the identification of growth areas within a backward area becomes a sub-strategy for clustering investments in sectors and areas of comparative advantage and thus providing the critical mass necessary to bring about a much bigger developmental impact over an area.

Summing up

To sum up, the planning for distressed pockets in a developing country calls for a sustained and systematic effort in planning and development. The critical components of such strategy will be (i) adopting a *typological approach* to impart greater area specificity to all programme ingredients, (ii) *initiating a transition growth process* to identify a dynamic policy programme package and to bring about a proper sequencing of the development process, and (iii) exploiting the *comparative advantage* theory for suitably establishing appropriate sectoral and spatial leads in development. What is important is that the various strands or elements implicit in these approaches must be rolled into a single policy package, operationalised into suitable programmes and projects and worked out in an appropriate sequence.

The conclusions that we have set above serve only as a pointer to an approach. How the job itself should be done in any specific backward area situation in the context of several constraints operating therein is still a problem, not amenable to any simple recipe or formula. It is here that the problem of spatial backwardness still poses enormous challenge to the planners in the developing countries!

APPENDIX 1

Variables used for the Identification of Backward Areas in India

1. Density of population per sq. km. of area.

2. Percentage of agricultural workers to total working force.
3. Gross values of output of foodgrains per head of rural population.
4. Gross value of output of non-foodgrains per head of rural population.
5. Gross value of output of all crops per head of rural population.
6. Percentage of total establishments using electricity to total number of establishments (manufacturing and repair).
7. Percentage of household establishments using electricity to total household establishments.
8. Percentage of non-household establishments using electricity to total non-household establishments.
9. Number of workers in registered factories per lakh of population.
10. Length of surfaced roads per 100 sq. kms. of area.
11. Length of surfaced roads per lakh of population.
12. Percentage of male literates to male population.
13. Percentage of female literates to female population.
14. Percentage of total literates to total population.

NOTES AND REFERENCES

1. While the State and the District are administrative areas, the region is somewhat elusive to define. No standard classification of regions exists. They are sometimes used to denote large divisions of the country (groups of States, better designated as zones) or spatial entities within a State for example, Telangana Region in Andhra Pradesh. The national sample survey follows a scheme of 58 regions delineated by grouping districts on some homogeneous criteria. Some studies have used these regions as an analytical frame. See Moonis Raza : "Levels of Regional Development in India", paper to the Indo-Soviet Symposium on Regional Development and National Planning, Tibilisi-Baku Oct. 1978.
2. The precision of the estimates of SDP vary from state to state depending on the type of data available as also their method of computation. The Central Statistical Organisation (Government of India) has now prepared a set of comparable estimates of SDP (as distinct from those prepared by the respective state authorities) which is now most useful for inter-state differential studies. But even here, a factor which affects the comparability between states is the variation in the purchasing power of the rupee between States at a given point of time. A further

complication when using time-series data on SDP is the differential inter-temporal movement in prices among States.

3. R.K. Sampath, Inter-State inequalities in income in India : 1951-71. *Indian Journal of Regional Science*, Vol. IX, No. 1, 1977. Also the papers presented to the 52nd Annual Conference of the Indian Economic Association, Dec. 26-28, 1969 published in the Special Number of the *Journal of the Indian Economic Association*.
4. In the beginning of the planning era, even though West Bengal and Maharashtra accounted for only 16.5 per cent of the total national population, their share in net national non-agricultural income was as high as 30.89 per cent that is almost twice their share in population. Over time, West Bengal's index of economic performance for non-agricultural sector declined from 202.74 in 1950-51 to 184.3 in 1960-61 and to 145.86 in 1970-71. The same trend is observed for Gujarat, Punjab and other States. For Maharashtra even though its index increased from 181.83 in 1950-51 to 191.56 in 1960-61, it declined to 154.78 in 1970-71. For non-agriculturally average and backward states, the indices show steady increase, for example, Andhra Pradesh, Bihar, Kerala, Mysore, Orissa and Rajasthan. Uttar Pradesh was an exception to this trend. (See Sampath R.K. : Ibid).
5. Ashok Mathur, *Regional Growth and Income Disparities in India*, Centre for Regional Development, Jawaharlal Nehru University 1978 (Mimeo).
6. B.N. Ganguli and B. Gupta Davendra; *Levels of living in India : An Inter-State profile*, S. Chand & Company, New Delhi, 1976.
7. Ibid, pp. 160.
8. S.C. Gulati : Dimensions of inter-district disparities, *Indian Journal of Regional Science*, Vol. IX, No. 2, 1977.
9. The works referred to here are the following :

(i) J.E. Schwartzberg : "Three approaches to mapping economic development in India"—*Annals of Assoc. Am. Geographers* 52, 1962 p. 455-68.

(ii) Asok Mitra : *Levels of Regional Development in India*, Part I-A (ii) Vol. 1 1961 Census of India.

(iii) Prakasa Rao, Sundaram, Ramachandran and Tiwari, Regional Disparities in India—paper to the Colloquium on Regional inequalities of development, IGU Commission Meetings, Rio de Janeiro, April 1971.

(iv) Biplab Das Gupta and Morris Jones W.H. *Patterns and trends in Indian Politics*, Allied Publishers (Pvt.) Ltd. New Delhi, 1975.

(v) M.N. Pal, Regional disparities in the level of development in India, *Indian Journal of Regional Science* Vol. VII, No. 1, 1975.

The Studies listed above have used mostly the data of the 50's and 60's.

10. Om P. Mathur, The problem of regional disparities : An analysis of Indian policies and programmes in *Growth Pole Strategy and Regional Development Planning in Asia*, United Nations Centre for Regional Development, Nagoya, 1976.
11. It should however be noted that this may not necessarily imply a cause-effect relationship.

12. T.N. Srinivasan and P.K. Bardhan : *Poverty and Income Distribution in India*, Calcutta 1974. Among the contributions in this volume, Chatterjee and Bhattacharya have presented estimates of the average consumption expenditure per capita in the rural and urban sectors of 15 States on the basis of eight NSS rounds which cover the decade 1957-58 to 1967-68. Vaidyanathan employs the NSS results for 1960-61 and calculates the "percentage of poor population" defined as with per capita consumption of less than Rs 11 per month, which was for that year roughly half the national average. Bhatty provides annual per capita consumption expenditure averages for 14 States based on an independently conducted enquiry in 1968-69 by the National Council for Applied Economic Research.

    Another study is by Moni Mukherjee who estimates the "density of the poor" in the different States of India. According to him, 'Density is expressed by the fraction of the lowest 10 per cent of households in the All-India NSS sample which is found within the boundaries of a particular State. If poverty were evenly distributed throughout the country, the lowest 10% of consumers at the national level would constitute one-tenth of the population of each State. According to Mukherjee's study, Orissa has the greatest rural density of the poor. Kerala has the worst score for the urban sector. At the other end of the spectrum are Himachal Pradesh with no rural poor and Assam with no urban poor. Moni Mukherjee, Size and areal distributions of the level of living in India, *Sankhya*, Series B, Vol 31, Parts 3 and 4, 1969.
13. P.D. Ojha : A configuration of Indian poverty—inequality and levels of living in A.J. Fonesca (ed) *Challenge of Indian Poverty*, Vikas, New Delhi, 1971.
14. E.W. da Costa : 'A portrait of Indian poverty' in Foresca : Ibid.
15. V.M. Dandekar and Rath Nilakantha : *Poverty in India*, Indian School of Political Economy, 1971.
16. According to the official version, 'Poverty line' is defined as follows : "Poverty in general connotes deficiency or insufficiency of acceptable material needs to the people. These needs have often been expressed in terms of a basket of food items satisfying certain desirable calorie requirements. Monthly per capita consumption expenditure which enables purchase of this basket defines the poverty level or line."—Planning Commission : 'Poverty level in five Year Plan—1978-83, A Review Note.' (Mimeo)
17. The concept of poverty as a statistical measure—'Poverty threshold—is sometimes criticised as it does not provide for recognising the degree of severity of poverty of particular people in areas. E.W. de Costa suggests a "poverty band" concept dividing the population into the poor, the destitute and the seriously destitute. See da Costa, E.W. A Portrait of Indian Poverty in A.J. Fonseca (ed), *Challenge of Indian Poverty*. Vikas Publications, New Delhi, 1971.

    For a critical analysis of the contemporary approaches to poverty, see C.T. Kurien, *Poverty, Planning and Social Transformation*, 1978.

18. V.M. Dandekar and Nilakantha Rath : *Poverty in India*, Indian School of Political Economy, 1971.
19. Mellor, *The New Economics of Growth*, Cornell 1977.
20. V.M. Dandekar and Nilakantha Rath : *Poverty in India*, Ibid pp. 129-130.
21. The data used in respect of the three variables are : (i) The census data (1971) for agricultural labourers converted into Percentage of Agricultural labourers to total agricultural workers. (ii) Rural unemployment rate in person days as measured by the NSS 25th Round for 1970-71. The data on weaker sections refers to consumption standards amongst rural labouring households and cultivators who fall within the lowest 10 per cent of the size distribution of land holdings.
22. Alice, Thorner Kerala : A strange case of poverty, *Purusartha*, Vol. III, Paris 1977, pp. 141-53.
23. Centre for Development Studies, Trivandrum 1975, *Poverty, Unemployment and Development Policy—A case study of selected issues with reference to Kerala*, United Nations, New York.
24. The most important of these is tapioca. NSS figures for consumption in Kerala of bananas and coconut oil also appear far too low in the light of the Food Balance Sheets, while the coconut kernel is completely absent from the NSS listing. According to CDS findings, the available supplies of food in Kerala should have yielded an average of 2337 calories per person per day.
25. The important among these are : State Domestic Product per capita, the compound growth rate per year of agricultural production, the wage rates for male agricultural workers and various other social indicators including literacy rates and death rates.
26. Apart from the above mentioned two anomalies in the picture of undevelopment thrown up by a region-wise analysis of poverty related variables, some anomalies have also cropped up on account of the way in which districts are grouped together in the NSS regions. A case is the inclusion of Kutch, possibly a backward district in North Gujarat which appears according to the NSS data as a developed region.
27. For a comprehensive review of the various statements in the successive Indian Five Year Plans, see Rao, Sundaram *et al*, Regional Disparities in India, paper to the colloquium on Regional Inequalities of Development, IGU Commission meetings, Rio de Janeiro, April 1971.
28. Draft Fifth Plan Chapter 14 p. 282, Planning Commission, New Delhi, 1973.
29. Draft Five Year Plan (1978-83).
30. For a critical account of the various area programmes, see K.V. Sundaram, Some recent trends in Regional Development in India, in R.P. Misra *et al* (ed.), *Regional Planning and National Development*, Vikas, New Delhi, 1977.
31. District Planning in India has been advocated in different periods with different emphases. For a full account of the changing concepts in

District Planning and India's experience in this direction, see K.V. Sundaram, Country Report Part I: Multi-level planning at the District and block levels in India. Paper to Workshop on Improving the Methods of Planning for Comprehensive Regional Development 16 May-12 June 1978, United Nations Centre for Regional Development, Nagoya.

32. "Growth with Social Justice" was advocated as the objective of planning in 1970. This led to special emphasis on the relatively underdeveloped areas and sections of the population and the introduction of a large number of centrally sponsored special area programmes. Since the change in Government, the objective has shifted to "Growth *for* social justice", and a new thrust on "area planniag for full employment" has been added. With this shift in emphasis, "Block level planning for full employment" has become important in planning at the sub-national level. An average development block in India covers about 100 villages, an area of about 600 sq. kms and population about 80,000-100,000. For more details on Block-level planning see K.V. Sundaram, Poverty alleviation in rural areas through employment creation—implications of Block level Planning and other methodologies", *Manpower Journal* Vol. XIV, No. 4, Jan-March 1979, New Delhi.

33. The Industrial Policy Resolution 1956 stated "It is one of the aims of national planning to ensure that these facilities (that is power, water supply and transport facilities) are steadily made available to areas which are at present lagging behind industrially or where there is greater need for providing opportunities for employment provided the location is otherwise suitable." The Industrial Policy Resolution amended in 1973 stipulated that "in the implementation of the licensing policy Government will ensure that licensing decisions conform to the growth profile of the plan and that techno-economic and social considerations such as economies of scale, appropriate technology balanced regional development and development of backward areas are fully reflected.

34. R.P. Misra and K.V. Sundaram, Growth foci as instruments of modernization in India, Part II in Kuklinski A.R. (ed.), *Regional Policies in Nigeria, India and Brazil*, Mouton, 1978.

(i) K.V. Sundaram, Spatial planning for a tribal region, Chapter 19 in *Urban and Regional Planning in India*, Vikas, New Delhi 1977.

(ii) Raza Moonis, Chattopadhyaya and Baudhayan, Regional development : The analytical frame, *Indian Journal of Regional Science* Vol. VII, No. 1, 1975.

(iii) B.K. Roy Burman, Social Processes in the Industrialisation of Rourkela" Census of India 1961, Monograph I.

(iv) N. Sen Gupta, *Destitute and Development : A Study of the Bauri Community in Bokaro Steel Region*, Concept Publishing Company, 1976.

35. B.D. Sharma, *Tribal Development, The Concept and the Frame*—Prachi Prakashan 1978, New Delhi, pp. 111-117.
36. Committee on Unemployment, Ministry of Labour and Rehabilitation : Report of the Working Group on Financial and Fiscal Measures —Government of India, November 1972, p. 40.
37. V. Venkatesan, "Appropriate Management for Integrated Rural Development—Some lessons from Maharashtra." Paper to the RVB Silver Jubilee Symposium, Delft, Holland, 12 Dec. 1977. Also Arvind Mafatlal, Economic Growth with Social Justice 1971 (quoted in Venkatesan's paper).

    Venkatesan has documented four projects viz : 1. Powerloom project in which a large number of unemployed handloom weavers were rehabilitated in a powerloom project consisting of 48 powerloom cooperative societies, a centralised pre-weaving unit and process house; 2. a dairy farm project in which the target group farmers were made owners of good milch cattle and were enabled to become efficient dairy farmers; 3. a ready made garments project; and 4. a leather complex.
38. A pilot research project in growth centres was launched in India during the Fouth Five Year Plan (1969-74) in 20 blocks.
39. The operational guidelines for Block level planning framed by the Planning Commission includes 'spatial planning for revised minimum needs and socio-economic infrastructure' as an essential ingredient of micro-level planning. See "Operational Guidelines for Block level-planning"—Planning Commission (Government of India) 1979. Also "Report of the Working Group on Block Level Planning", Planning Commission (Government of India) 1978.
40. The various evaluations and reviews of the programme have brought out its limitations and deficiencies. The more important among them are : difficulty in identifying appropriate beneficiaries; confinement of benefits to the better-off among the rural poor; insufficient extent of subsidization; cumbersome procedures of credit institutions especially commercial banks; weaknesses in the working of co-operatives; rigidity in the schematic pattern of the programme and in central control; exclusion of infrastructure development and farmers' training in the programme package and lacunae in the personnel management emanating from frequent transfers, delays in filling up of posts, weaknesses in supervision, absence of regular monitoring; and lack of competence and apathy of project-personnel at contact level.

See (i) S.M. Pandey and J.S. Sodhi, *Developing the Rural Poor—A study of marginal farmers and agricultural labourers development agency in Delhi*, S.R.C. for Industrial Relations 1977.

(ii) R.C. Arora, Development of Agriculture and allied sectors : Integrated area approach, New Delhi, 1976.

(iii) V.R. Gaikwad : Participation of Rural institutions and target groups in rural development programme, Planning and Management. Paper presented to the Round Table on 'Adaptation of

Administration to Rural Development', Government of India and ESCAP, 16-18 August, 1978, New Delhi.

(iv) Dwarkinath, *et al* : Potentialities of marginal farmers for agricultural development—University of Agricultural Sciences, Bangalore, Technical series, No. 7.

41. The Task Force on Integrated Rural Development appointed by the Planning Commission, Government of India in 1971 under the chairmanship of B.S. Minhas. The Task Force submitted its report in June 1973.
42. The Task Force has recognised three broad categories of areas based on the quantum of rainfall. In each area, however, further sub-divisions characterised by distinct land use and cropping patterns exist.
43. D.T. Lakdawala : Draft Plan : Forces of Growth. S.V. Desai Memorial Lecture, Ahmedabad, 24.3.1979.
44. V.S. Vyas, Some aspects of structural change in Indian Agriculture, —Presidential Address, Indian Society of Agricultural Economics, 18.1.1978.
45. Pranab K. Bardhan, India—Country experience in Holis Chenery *et al. Redistribution with Growth*, Oxford University Press, 1974.
46. The unit of identification was the Taluka (District). Five different exercises were undertaken. One of the exercises was based on the ranking method which was not considered useful. The other exercises were based on construction of indices of development worked out with State average taken as 100. The indicators were varied in the different exercises, though 5 indicators were common to all of them. The cut-off points for determining backwardness were decided by the respective Planning and Development Committees. 52 talukas out of 80 in coastal Andhra, 68 out of 73 in Telangana and all the 42 talukas in Rayalaseema were declared backward. Thus 162 talukas in the State constituting 72.6 per cent of the population of the State was declared backward. See Narayana D.L. : Problems of Identification and Policy for the Development of backward areas : A case study of Andhra Pradesh —paper to the Seminar on the Development of Backward Areas, Indian Economic Association, Patna, 1979.
47. Planning Commission (Government of India) 1979: Operational Guidelines for Block-level Planning.
48. Anand Sarup, Case of the pilot project for six Development Blocks of Ghazipur-Ballia districts in Country Report on India—Part II—Workshop on Improving the Methods of Planning for Comprehensive Regional Development, 16 May-12 June 1978. United Nations Centre for Regional Development, Nagoya (Japan).
49. The programme content in the Action Plan included among others, old age pension, land allotment, milch cattle units, goats and sheep units, land improvement, purchase of bullocks, agricultural inputs, hand operated loom, shoe-making, pottery, bangle making, sewing machines, assistance for running petty shops etc.

50. Planning Commission (Government of India): Report on a quick concurrent evaluation study of the working of the Antyodaya programme in Rajasthan, January 1979.

51. A criticism is voiced sometimes that an area development plan leads only to "place prosperity" and not to "people's prosperity". This criticism has to be taken seriously in the developing countries, for it has stemmed from the character of the area development programmes implemented in most developed countries which are generally in the nature of infrastructure programmes. In the developing countries, on the other hand, planning has to be from the human end.

52. The amount of subsidy (Rs. 1500) mentioned here is a rough approximation. The exact amount of subsidy and credit will depend on the carefully worked out requirements of each scheme under which he is helped.

53. This will include: the inherited background factors of development, the economic-geographic features peculiar to the area, the natural resources endowment and the indigenous entrepreneurship and skills in the area.

54. The directions of structural change for the development of small tribal communities, for example, the Abhujmar area in Bastar district have been outlined by Sharma. See B.D. Sharma, Development of Small Tribal Communities, Chapter V in *Tribal Development: The Concept and the Frame*, Prachi Prakashan, 1978.

CHAPTER 10

LESSONS OF EXPERIENCE
The Task and the Means

> Wherever we look, we come to the same conclusion: in the Third World, the White man has started something for which the term "development" is not an accurate description, but a euphemism. The continuation of what was and what is does not by a long chalk add up to a future. Nothing swings back into line............ It will require a planned, purposeful joint international intervention if this course is to be altered in time. The Third World has not much time left.
>
> —Erhard Eppler in "*Not Much Time for the Third World*"

> We tremble to survey
> The growing labours of the lengthen'd way
>
> —Alexander Pope.

The previous discussion on the theory and practice, highlighting the current policy and programme measures aimed at the closure of the various types of inequalities in the developing countries, has yielded considerable insights. This chapter spells out some of the main conclusions which can be drawn from this analysis and seeks to identify the elements of a strategy that can perhaps work in the context in which the developing countries are placed. Undoubtedly, the experimentations in terms of strategies and methods for achieving a more even development pursued by the various governments, (recorded in the case studies), may not yet provide definite clues. The diversity of the responses explored and deployed to meet the different problem contexts, as well as the relatively short duration of some of these experiences, render our problem of drawing valid generalisations somewhat difficult, nay even controversial. The author also subscribes to the view that any development strategy devised for stimulating growth and efficiency and of narrowing inter-regional disparities must be essentially country-related and in line with the national goals. Further backward area development cannot be viewed as another specialisation as commonly understood, because its subject matter cannot be isolated from development planning in its totality. Nevertheless, there are certain unique and specialised strategies, methods and policies which individual countries have evolved as well as certain commonalities in their experiences and approaches which provide insightful and retrospective material for our interpretative analysis. Based on such analysis, I have ventured to outline some fundamental postulates to the formulation of strategies for spatial equalisation and backward area development at the end of this chapter.

To begin with, let us briefly recapitulate the major findings of the case studies.

THE PHILIPPINES

In the Philippines, the socio-economic development of the various regions within the country became an explicit objective in the enunciation of government policy since 1972, more than 25 years after the country became independent. Historically, the spatial structure of underdevelopment consisted of an excessive concentration of development in the island of Luzon, particularly Manila and its environs and a few urban centres in the other islands, leaving glaring developmental gaps in the rest of the country.

Among the strategy and policy measures pursued by the government to generate a reversal of the regional growth trends, the following stand out as significant :

(1) Indirect methods, particularly industrial incentives, to curb the polarized development in Manila and its environs and to assist industries to move to other areas;

(2) formalized regional development planning through regionalization of the country and creating an instrument called the Regional Development Council in each of the thirteen regions in the country as an administrative vehicle for regional planning;

(3) some limited Integrated Area Development programmes under special and often times ad-hoc arrangements; and

(4) limited local autonomy measures for nurturing the traditional village organizations—the barangay councils and assemblies.

Among the above measures, the indirect methods did not result in any significant changes. The other three are direct methods, all nationally conceived and centrally directed. They are intended to inject a spatial dimension into the national planning effort. Thus along with the vertical development approaches, the emergence of horizontal development approaches with the fundamental objective of regional and local (area) development "to explore the fullest potential of each region while reducing geographic inequalities in development and development opportunities" is clearly evident. Problems of coordination, planning capacity (especially paucity of planners and incapacity of local governments), adaptation of the administrative structures and insufficiency of funds have led to limited scale impacts of these measures. Undoubtedly, these initiatives take a long time to gestate. In the meantime, several innovations are being made. Regional budgeting and their integration with the sectoral budget at the national level to ensure that each region gets its priority projects implemented has been accomplished. Several administrative models are being tested. The area development programmes that have been launched in selected areas and the local development efforts through the barangay councils and assemblies are particularly of interest to us in the present context, as they shed light on the viable strategies for the development of backward areas. From these experiments, it is clear that development efforts targetted on the rural sector and the small farmer and focussing on delimited geographic areas have great potential. Also their success depends on the institutionalisation of the rural poor and participatory approaches. The IAD programme of the Philippines has these features. They also contain ingredients of policies to improve the assets position of tenant farmers and landless tenants, provision of much greater direct support to productive activities upon which the poor are heavily dependent and provision of the basic minimum needs to the rural population. However, these programmes are of relatively recent origin. From the available evidence, it is difficult to conclude whether spatial inequalities are decreasing or increasing. However some indications lead to an inference that the proportion of poor families are increasing.

MALAYSIA

Although Malaysia gained independence much later (1957) than the other countries selected for case studies in this volume, the country's development path shows that there has been a sharpened focus on spatial imbalances and regional issues. The so-called regional problem in Malaysia is not merely a spatial problem, but a racial, religious and cultural one too. The minimization of interregional socio-economic development imbalances is enshrined as a definite policy objective in the country's development plans since 1971. The developed regions of the country are along the west coast of peninsular Malaysia with urban growth and activity concentration in the Klang Valley. This may be recognized as the central core developed region of the country. In the interior and in the Sabah and Sarawak Islands exist vast virgin lands with great potential for development, constituting a resource frontier. The poor or lagging regions are in the states of Kelantan and Trengganau of the east coast and also in an area around Penang in the north. There is heavy federal control on financing and allocation of developments and planning is therefore centrally directed and controlled. The planning leverage at the local and State level is limited. Since there are plenty of virgin lands in the interior, development is taking place through frontier expansion. Land development, infrastructural development and a few sporadic comprehensive regional development efforts are spearheading this interior backward area development strategy in Malaysia. These frontier development experiments have yielded some valuable insights into the factors that ensure sustained development in the peculiar situation of Malaysia. They have underlined the need for a carefully conceived institutional mechanism to carry out the various facets of development, the attention to the human resource component and grassroots level participation. Among the many experiments made are some success stories for example, the FELDA model of land development, the Muda

Irrigation Project and the Pahang Tenggara regional model of development. Simultaneous with frontier development, financial and fiscal incentives are being given to prospective entrepreneurs to promote dispersal of industries; vocational training and education are being imparted for the improvement of skills in general and for increased Malay participation in modern sectors in particular. All these constitute the various dimensions of a top-down process of development. There is tacit recognition for bottom-up approaches in the development of the rural sector. But these are yet to be articulated effectively. The Operations Room and the Red Book techniques constitute important lessons in implementation methodologies.

INDONESIA

Indonesia represents a plenum vacuum pattern of development with developments concentrated in Jawa, Sumatra, and Kalimantan (Jawa constituting the core) and the outer islands remaining in a raw material extraction and export stage of development. The income from the outer islands was being siphoned to add to the prosperity of the foreign companies, State enterprises and Indonesian business interests residing outside the producing provinces. Attempts are being made to reverse this kind of development process. Regional policies have received explicit consideration since the launching of the Second Five Year Plan (1974-79). Regional development problems are being sought to be tackled in to major directions (a) through population redistribution and settlement, by moving labour to jobs through planned trans-migration programmes and (b) through rural development which is being promoted through Presidential grants-in-aid projects at various levels.

Experiences with transmigration programmes have yielded considerable insights into the factors that contribute to their success. It has been shown that the various development elements must fall in their places in appropriate sequences and in right time; that meticulous planning involving total comprehension of the development effort is needed; and that appropriate organizational, institutional and development administrative structures are to be devised. The rural development programmes are being sought to be promoted through development subsidies (unit or per capita grants) by the central government to the second and lower levels. Thus there are the Provincial Subsidy (Peoyer Inpress Propinsi), the subsidy at the Kabupaten (district) level, called Inpres Kabupaten and the village subsidy. There are two aspects of the rural development programme in Indonesia which are worth mentioning. One is a village progression concept implied in the rural development effort which envisages a three-stage transformation of villages—From Swadaya through Swakarya and Swasambeda to Pancasila village. This transformation is sought to be realised within 25 years from 1974. In order to accelerate this transformation, an integrated area development approach (called UDKP) with Kecamatan (subdistrict) as the unit area is being experimented. All these, however, have resulted only in an executive type of decentralisation—otherwise called deconcentration—which implies central initiative and direction in all matters primarily through bureaucratic channels.

The Indcnesian economic development policies are not, as yet, sufficiently informed by the principles of redistribution with growth, structural reform of society and anti-poverty strategies. The development problems faced in the Indonesian context are : ineffective institutional framework, scarcity of grass-roots local leadership, inefficient coordination of development efforts, inadequate administrative machinery, lack of planning capabilities, weakness in implementation and lack of efficient control of the implementation activities of the development programmes.

THAILAND

Economic planning has had a relatively late start (1961) in Thailand. The concern for sub-national (regional) development came to be voiced since the Third Five Year Plan (1972-75) and became more explicit and articulate in the Fourth Plan.

Economic activities in the country are concentrated in the central region in which Bangkok Metropolitan area is located. The problem of disparities is one of a primate city and its underdeveloped hinterland comprising all other regions of the country. Although free from colonial rule during the last 200 years, the spatial structure of underdevelopment in the country has all the characteristics of colonialism with an externally oriented economy and with its interior characterised by spatial penetration rather than spatial development.

So far as development of lagging regions is concerned, there are some experiences worth noting, particularly in the north-east, north and south regions of the country.

The development of the north-east region, which is considered to be the most underdeveloped region, received some attention since the 50s. The pressure for this development came in the wake of insurgent activities linked with the events in neighbouring Indo-China. This took the form of the development of certain vital infrastructure, particularly transport and irrigation, all of which were geared to security rather than development. The rural development schemes were more in the nature of propaganda stressing the evils of communism. Thus despite the Thai government treating the north-east as a unique problem, the impact was limited.

For the north, which is another lagging region of Thailand and which has also a sizable component of tribesmen in its population composition, a study team has favoured concentration of effort along an identified growth area accompanied by a strategy of selective closure of this area in the first instance and gradually drawing this area into the normal developmental stream.

The south, with its economy based on primary produce—rice, tin and rubber, owes its underdevelopment to the low level of industrialization and fluctuations in international commodity prices. A plan prepared for this region favours a concept of balanced regional development based on concentration of effort in the most exploitable resources of each province or area. It also favours indirect policies such as policies designed to accelerate urbanisation and those of developing new agricultural areas.

In brief, Thailand's experience suggests a three-point strategy to tackle the problem of bringing about a more balanced and integrated economy. These include :

(i) a strategy for the intensification and diversification of the economy for improving the regional industrial mix and export base;

(ii) a strategy for decentralised urbanisation; and
(iii) a strategy for decentralised infrastructural and social services to the rural areas.

The rural development objectives in Thailand are sought to be achieved through extension strategies, which do not reach the poor. There is an outright grant to Changwats (Provinces)—called the Changwat Subsidy Fund—introduced recently (1979) which is similar to the INPRES Grants in Indonesia, but on the whole, local level planning and development are not much advanced in Thailand and decentralisation processes still have a long way to go, feced as they are with a number of obstacles such as local power structure, dependency syndrome and bureaucratic obstructions.

INDIA

Since attaining independence in 1947, India has been pursuing the goals of planned development, aiming at balances in various spheres, through successive Five Year Plans. However, it is only since the 4th Five Year Plan (1969-74) that some direct Central interventions aimed at spatially oriented development for certain types of backwardness in the country assumed significance. These measures have been continued and even enlarged during the 5th and 6th Five Year Plans.

The spatial structure of underdevelopment in colonial India consisted of a metropolitan economy concentrating on Port enclaves extending its tentacles into selected plantation and mining areas along with a vast rural hinterland of primary produce. The contours of this spatial structure are being modified. The relatively less developed areas are in the east, centre and middle south parts of the country. With 46 per cent of population estimated as living below the poverty line and with the lowest 40 per cent of the population strata sharing less than one-fifth of the nation's GNP, the inter-personal income inequalities are high.

In India's repertoire of policies and strategies aiming at balanced regional development, there are a number of indirect and direct measures which are of great interest. These include—

(i) an inter-regional allocation policy for the distribution of Central Assistance funds to the State Governments which is governed by a formula tilted in favour of backward areas. In recent years, by including the inverse of state income per capita in this formula, more refinements have been introduced;

(ii) capital and technology transfers to backward areas;

(iii) incentive policies designed to direct investments to the industrially backward districts;

(iv) growth centre policies;

(v) the adoption of some area specific, problem oriented sectoral schemes;

(vi) a sub-plan approach for undertaking comprehensive area planning in the country for some special problem areas (Action plans based on area development approach);

(vii) a basic needs strategy oriented towards the provision of minimum needs, so that the disadvantaged areas and groups may achieve parity with others in terms of social consumption;

(viii) asset redistribution policies;

(ix) regional planning approach for inter-State regions and special problem areas; and

(x) comprehensive local level planning focussed on the District and the Block with employment and other multiple objectives for local area development.

The policy of capital and technology transfers by locating large industrial complexes in backward areas, by and large, has not succeeded bacause of the failure to carefully plan for the linkage effects of such investments by following up this strategy with a series of other sectoral and area development strategies. These experiments only led to certain islands of development with no spread effects on the countryside. The growth centre policies in rural areas have indicated that they should not be pursued in isolation but along with properly conceived local area plans aiming at the development of local potentials. Also such policies in resource-rich tribal regions must be pursued along with a policy of cautious incrementalism and comprehensive planning. The working of incentive schemes for industrial development have brought out their general inadequacy in stimulating the development of backward areas, as local entrepreneurship from these areas fail to come forward to avail of these incentives. The lessons learnt from this experience are—that these incentives should be more selective in terms of specific industries and locations and also that such incentives should be designed over a wide spectrum of economic activities. The sub-plan approach which combines some features of a Spatial Budgeting approach and a transition growth strategy seems to hold great potentials for the development of backward areas. As practised in the Indian context, they need considerable refinements. The asset redistribution policies like the land reforms have proceeded at very slow pace. The regional approach followed in some inter-State regions have had limited impact. Comprehensive local level planning focussed on the district and the block with employment and other multiple objectives is making some headway. Apart from the methodological problems, two other factors that are in the nature of constraints are : (a) the problem of coordination and (b) planning capacity, which includes not merely inadequacy of trained planners but also in capacity of local governments and the problem of insufficiency of funds for local level planning.

The eclectic approaches to the problem of inter-regional disparities attempted in India point to several important lessons. Some of these are—

(1) The development approach aimed at backward area development should be characterised by (i) selectivity; (ii) specifity or focus; and (iii) dynamism or flexibility.

(2) The strategy for backward area development should be engineered over a long period of time during which development proceeds in a chain of see-saw relationships through an intervention process alternating between selective differentiation and integration approaches.

(3) The development of backward areas could be promoted in a significant manner by increasing labour productivity, improving existing capacities, making deliberate

efforts to capture the various interlinkages and externalities in development and thus improving the management of development for better performance.

(4) The backward area problem cannot be solved by adopting a blanket approach. Each type of backward area has to be identified and appropriate strategies devised for each case. A typological approach to the problem is indicated.

(5) The development of underdevelopment has to be conceived as a time-bound process of transition progressing in stages according to a pre-conceived perspective plan.

Major conclusions emerging from the case studies

The countries discussed fall into varying categories in terms of population, GNP, per capita GNP, levels of development, degree of inter-regional disparities, extent of poverty, income distribution and the degree of decentralisation. The table 10.1 reflects these variations.

In terms of population, India and Indonesia lead, followed by the Philippines and Thailand in the second group and Malaysia falling in a third group. In terms of GNP, however, the position is reverse. In terms of growth rate of GDP per capita, Thailand leads followed by the Philippines, Indonesia and Malaysia with India occupying the last position. In terms of regional disparity ratio, India shows a comparatively better position. But as the bases of comparison are not uniform between the countries, it would be misleading to draw valid conclusions in this regard. In terms of poverty levels, Malaysia has the lowest percentage of population below the poverty line and India and Indonesia, the highest; the Philippines and Thailand come as an intermediate group. The same position is reflected in income distribution also. In the scale of decentralisation, all the countries are in the de-concentration stage only. There is, however, keen desire in all these countries to move towards the goal of decentralisation, as is seen from their plan documents. But as of today, only top-down approaches to planning prevail. In a country like India, there is a move to establish multi-level planning

TABLE 10.1 : **Some Indicators of Development**

Country	Population (Million) 1975	GNP group	Growth rate of GDP per capita 1975	Regional disparity ratio i.e. the ratio of GDP per capita between the highest & lowest state/province in the country	% of people below poverty line † 1975	Share in GNP of lowest 40% of population † 1975
1	2	3	4	5	6	7
Indonesia	130.0	A	2.2	1 : 12.3*	59	16.1
India	599.4	A	1.8	1 : 2.7**	46	17.0
Philippines	42.5	B	2.8	1 : 3.7***	33	11.6
Thailand	41.6	B	3.5	1 : 3.7+	32	11.5
Malaysia	12.2	C	2.1	1 : 3.6++	12	11.1

NOTE : A—Under $ 350 (CP).
B—Between $ 350 and 750.
C—Greater than $ 750.
*—For 1972.
**—Average between 1970-73.
***—For 1974.
+—For 1973.
++—For 1970.

†*Source :* Growth and poverty in developing countries, World Bank Staff Working Paper No. 309 (Revised), May 1979 Tables 1 and 2 for all columns except 5. Figures in column 5 are compiled by author from various sources.

procedures. States within the country have responded differentially to such decentralised planning methods.

The spatial pattern of development

The spatial configuration of development in the countries analysed here is undergoing many changes under the impact of planned economic development. From the stand-point of the spatial structure, these countries may be grouped into two broad categories—(1) Countries with low density of population associated with considerable virgin land within their territories, which are still engaged in frontier development, seeking major shifts in their geographical division of labour; and (2) countries with relatively higher density of population seeking intensification and diversification of their economies and going in for various adjustments through multi-level planning efforts. Malaysia and Indonesia are typical examples of the first type, while India is typical of the second group of countries. The Philippines and Thailand seem to have a mixed bag of characteristics, that is, partaking characteristics of both. It must be noted that within the limits of the above generalization, the individual countries themselves have their own historically unique configurations of development which has been referred to in the case studies.

All the countries studied (with the exception of Thailand) have emerged from colonial dependence and show features of an unbalanced and unintegrated economy with gross spatial inequalities. The spatial pattern of colonialism in these countries consisted of a metropolitan economy centred on one or more major ports in the country, extending its tentacles into plantation and mining enclaves in the interior and with the rest of the country constituting a vast rural hinterland of primary produce, as in the Figure* below :

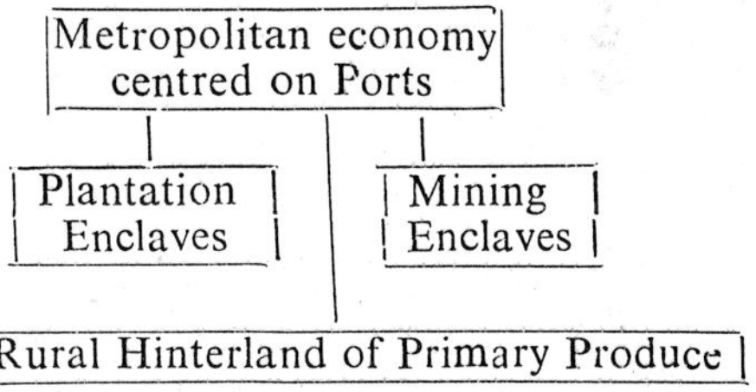

This structure is changing, but slowly, after national governments came into existence in these countries. During the course of this transformation, capitalist interests in these countries (including multi-national companies) have tended to

* Adapted from Moonis Raza, Atiya Habib and Amitab Kundu : Spatial Organisation and Urbanisation in India. A case study of underdevelopment in R.P. Misra and K.V. Sundaram, *Rural Area Development : Perspectives and Approaches*, Sterling, New Delhi, 1979.

follow a direction of development which has led to the persistence of these patterns in a slightly modified form—often referred to as a structure of internal colonialism. It is in this context that some scholars have argued that what is needed for these countries in the first instance is not so much increase in the growth rates, but a change of direction. In the context of the existing pattern of spatial inequalities in these countries, and in view of the fact that the path of development towards growth combined with equity, will necessarily be a long and winding one, this author also firmly believes that what should be realistically attempted is to trigger off a process of change in the right direction. There seems to be considerable agreement in principle on this point in these countries, as revealed in their recent Plan documents. What seems necessary is more clarity in our thinking, as to how to trigger off such a process of change.

The nature of the inequality problem differs from country to country. For example, in Malaysia the spatial inequality problem is compounded by ethnic disparities. Geographically, the developed, the lagging and the frontier areas are distinct. Indonesia's development follows a plenum and vacuum pattern. In Thailand, it is a case of primate city and its hinterland characterised by backwash effects. In the Philippines also, a similar pattern prevails. In India, the areas of underdevelopment are concentrated in the east, centre and middle south parts of the country. Thus the present spatial pattern of development in all these countries reveals a patchwork—characterized by a spatial penetration of development here and there in the interior rather than an integrated pattern of spatial development.

In all these countries, economic planning has been a comparatively recent phenomenon, perhaps post-war or post-independent. Their planning documents show that regional policies are also becoming explicit in varying degrees, especially since the 70s. It seems reasonable to make a historical generalisation that in the early stages of a nation's development, macro-level issues become heady, muscular and aggressive and the growth objective claims foremost attention. As a nation develops, more and more attention is focussed on the developmental issues of the sub-national areas. In the beginning, only ad hoc arrangements are pursued, which are not motivated by any wish to introduce a general programme of regional and local development and still less by any desire to plan fully for the national space economy. In course of time, however, as a response to many regional problems and spatial dilemmas, regional policies tend to evolve. This appears to be the nature of developmental sequence that comes about, as may be generalised from the trends witnessed in these five countries. This sequence appears to be somewhat in contrast to the historical experience in many of the developed countries (for example, Britain) where local planning preceded national planning. As a result, there evolved at the local and regional levels, professional bodies which demanded solutions and actions on spatial issues[1]. By contrast, in the developing countries, planning has been introduced from the top-down.

In the works of planners such as Friedmann, Alonso, Rodwin and Berry[2] it has been argued that spatial planning is essential at an early period of economic development rather than being a later and optional planning extra. Friedmann has also argued that "transitional societies are clearly most directly concerned with regional organisation partly because of the spatial shifts involved in moving from an agrarian to an industrial economy, and partly because a large portion of their potential resources are still unutilised."

The above arguments indicate clearly that in the context of the late awakening to regional problems, the developing countries have to make rapid strides to catch up with the lost time and opportunity in order to put their space economy in order.

The policy instruments used in the various countries to bring about a more even development range from indirect to direct and from selective to comprehensive measures. They include—

(1) Policies for inter-regional allocation of resources;
(2) Capital and technology transfers to the relatively underdeveloped areas;
(3) Incentive policies of various kinds designed to direct investments with varying degrees of locational specificity

(activity) spheres of economic activity, for example, industry, agriculture, transport etc;

(4) Growth Centre Policies;
(5) Target Group-oriented policies;
(6) Policies oriented to the preparation of action plans in selected problem areas based on the area development approach;
(7) The basic needs strategy;
(8) Asset redistribution policies;
(9) Regional planning policies; and
(10) Local level planning policies.

The country experiences have brought out the many constraints faced during the implementation of the above policies. They have also shown that direct policies are more productive of results than indirect policies.

The studies have brought out very clearly that the nature of under-development in the developing countries is not merely one of regional income disparities but also wide disparities between and within social groups and also disparities in natural resources development. Thus geographical inequalities, developmental inequalities and socio-economic inequalities have all to be tackled together in the developing countries. In this context, it is clear that no single set of policy instruments will be adequate.

The broad approach followed in these countries is to bring about certain adjustments to make the geographical division of labour conform to the basic geographical realities. Geographical inequalities within a country are being sought to be tackled through various methods, depending on the complexion of the problem. These include, land settlement and frontier development, encouraging internal migration, mitigating environmental constraints through the application of science and technology and intensification and diversification of the development effort. Development inequalities are constantly being tackled through inter-sectoral and inter-regional allocation policies involving continuous adjustments to bring about the necessary correctives. In order to tackle the problem of socio-economic inequalities, the above strategies are combined with several protective as well as anti-poverty strategies focussed on specific target-groups.

In this manner, the problem of spatial inequalities is being tackled through numerous policies and strategies in various countries. The differential emphasis given to the various types of policy instruments will depend on the geography, the inter-regional levels in development as well as the pace of development effort in the countries concerned. Those countries which are still wrestling with problems of land settlement, frontier development and redistribution of their population are not able to devote much attention to sub-national problems of adjustments/readjustments and distributive and decentralisation policies. On the other hand, in the countries, where there are no open spaces and there is little scope for any frontier development and where a certain pattern of geographical division of labour has already come to stay, increasing attention is being paid to the intensification and diversification of the development effort, to the problems of adjustment/re-adjustment and to the distributive and decentralization policies. This seems to be the trend of development affairs in the developing countries today.

It is always difficult to generalize the successes and failures of planning efforts, particularly when multinational comparisons are involved. This is because too many independent variables affect the outcomes and the outcomes themselves are viewed differently in the different countries according to the value premises adopted. What is needed is an appropriate policy package, the content of which will naturally vary for each country. Some of the sources of variations for these policy patterns are as follows :

(a) What is the magnitude of inter-regional disparities in the country?

(b) Is the country politically dependent, newly independent or completely autonomous?

(c) What are the pre-existing factors that shape the impact of the forces of regional economic development?

(d) Do pressures for closing the inter-regional gap come from the internal implications of a value system that is from the innate desire to bring about an ideal geographical division of labour in the country through the exploitation of comparative advantage of regions; or other economic pressures as in the case of population pressures on the land (economic) or political pressures arising from a threat to national unity or economic and political pressures combined?

(e) Variations in the path toward development. Is there any attempt at sequencing of development along some preconceived lines? If so, what are the major elements that differentiate one stage from the other?

(f) What is the nature and the degree of decentralisation envisaged in the country keeping in view the current level of development and available capacities for development?

The above sources of variations render it virtually impossible to establish hard and fast generalisations as to the most desirable policies and strategies for the closure of the inter-regional gap.

A point emerging from the case studies is that a policy of merely pumping in resources in the underdeveloped area will not automatically induce even development. This is because in the underdeveloped areas, there is both a resource constraint as well as a demand constraint. In order to take care of this double constraint, sustained investment effort must go hand in hand with improving the economic conditions of the masses, particularly the wage levels of the poor, thus enlarging the consumer goods market. In densely populated countries like India, it may call for a strategy of sustained public sector investments in labour-intensive projects.

A major problem emerging from the case studies is the wide divergence noticed between policies, plans and performance. The reasons for such divergence appear to be primarily the lack of planning capabilities to translate the policies into effective plans, programmes and projects and the lack of administrative capabilities to carry out the plans and programmes effectively into action. Translation of policies into programmes and their implementation at the local and regional levels have always been a struggle with reality. Some of the difficulties atleast arise from the fact that planning operates at several levels of the Governmental hierarchy and effective multi-level planning procedures have not yet been evolved in these countries. For instance, even an apparently simple operation like regionalising the national budget causes conflicts both within bureaucracy and political levels. In the case of implementation of programmes at the local and regional levels, the problem essentially is one of lack of coordination among the line-agencies, particularly when horizontal integration is desired. In this context, coordination implies many things; not merely administrative coordination, but others like methodological coordination, coordination of the activities of non-govermental interest groups as well as ex-ante and ex-post gap coordination. It cannot be over-emphasised that much would depend on the institutional machinery devised for area planning and implementation. Some insights into the requirements of such machinery have been afforded in the case study discussions relating to the mechanism for area development (for example, Muda Irrigation Project in Malaysia, Transmigration programmes of Indonesia etc.).

Regional and local development undoubtedly stands out as the major plank of the strategy for achieving a measure of inter-regional equalisation. If this is to come about in a realistic manner, the Governments of these countries have to make determined attempts towards decentralised planning and implementation. In the context of a highly hierarchical and centralised system of administration prevailing in these countries, this has proved to be a formidable problem. The case studies have shown clearly that these countries are far from achieving the decentralisation goal. What has been attempted in most countries is some form of executive decentralisation (better called de-concentration). Even in India, where the decentralisation trends have gone a step ahead of the other countries, the achievements have been quite uneven among the various States within the country and integration between democratically elected local bodies and the bureaucratic developmental administration has been far from satisfactory. The bottom-up approaches in planning are bruited about, more to serve political ends; they have not taken roots in the system. In practice, they have taken the form of half-hearted measures (for example, the presidential INPRES grants in Indonesia, the Changwat subsidies in Thailand and the Un-tied Grants for local development in India). The case studies clearly show that strong political will for decentralised development and concerted measures for local and regional development are yet to come about in these countries. Some success stories have been experienced here and there; but it has not been possible to replicate them on a country-wide scale on account of various constraints (for example, financial, personnel and institutional constraints as well as lack of political will).

The removal of spatial imbalances or inequalities is a conceptual purity. It is only a mental construct, which cannot be found anywhere in reality. Development is always dynamic and shifting in terms of space, sectors or community. In fact in Hirschman's analysis, such imbalances provide the trigger for constant change. As he says, "development moves only through a chain of dis-equilibria." The movement towards spatial equilibrium, even somewhere near this goal is a long-term process for the developing countries, which are faced with many constraints of a financial, personnel and institutional nature. In a realistic, social and political context, the developing countries can only hope to achieve balance within what may be called certain tolerance limits. These tolerance limits are difficult to define. They refer to a fuzzy frontier, within whose boundaries the degree of discomfort created by development is a minimum and beyond which it leads to a problem of serious regional conflicts, which often have their political overtones. Obviously, the gap between what ought to be done and what is possible, is one of the most distressing experiences of all in the field of development policy.

If the above hypothesis is correct, one may conclude that during the process of development, balances of various kinds are to be constantly aimed by the National Governments. It has to be an iterative process in which one has to incessantly march towards the ultimate goals in a series of stages and steps. This is inevitable. In the context of the developing countries, the attainment of a perfect equilibrium even in parts, (either in sectors or in space) cannot be achieved all at once. It has to be pursued in a highly selective manner through in a process initiation by integration. This is somewhat akin to the appearance and reappearance of various characters in different combinations in a drama, as the scene shifts each time. The point that I am making is that the Development Drama that we are witnessing in the developing countries today is one replete with alternate interventions, representing each time, a struggle for a certain type of equilibrium only.

The development process understood in this manner, is one continuous series of adjustments, and although certain short-term and medium term goals may be attained during this process, the long-term goals are seldom within reach. They seem to be always receding. In other words, there is nothing like reaching a stage of perfect equilibrium or achieving a total convergence of disparities.

Williamson's hypothesis of the convergence relationship between per capita income and regional disparities assuming the shape of an inverted 'U', is in this author's opinion an abstraction and an unreality. So far as the developing countries are concerned, it seems to this author that its attainment over a normal temporal span, representing one or even two human generations, has to be regarded as a distant possibility.

The empirical study of the problem of regional imbalances in India shows that there was a tendency to convergence in the first decade of planned development and since then, the trends towards divergence are noticeable. It seems that in the dynamic situation obtaining in the developing countries, there will be many jerks during the development process indicating several back and forth shifts. In other words, what one may witness will be a repeated see-saw of relationships towards convergence and divergence. In the complexity of the situation prevailing in the developing contries, generalizations about the temporal behaviour in spatial inequalies seem to be difficult.

In the effort made by governments to combat the problem of spatial inequalities, there are certain policies that operate over the entire national space and which over time, set in train certain processes leading to the overall convergence of development disparities. Undoubtedly, this will be a long term process. In order to accelerate this process, different governments have devised certain other policies, which operate differentially in space and among target groups in society. These are the selective policies pursued for the backward regions/areas/classes and for the overgrown metropolitan areas in the country. Regional imbalances comprising of under/over development are two sides of the same coin. In the present work, the emphasis is on the policies relating to underdevelopment. While we cannot be quite dogmatic about the particular policies and strategies to be followed by the countries, yet it is possible to state some fundamental postulates in the nature of broad guidelines to the formulation of strategies and policies leading to spatial equalization and backward area development. We may therefore turn our attention to these fundamental postulates.

Some fundamental postulates to the formulation of strategies for spatial equalization and backward area development

In this long haul of discussion, cutting across the approaches of many disciplines, we have waded through a turgid literature of numerous concepts and approaches, often of a conflicting nature, and attempted to take stock of the several theoretical frameworks that exist for guidance in designing and administering programmes for the amelioration of backwardness in the developing countries. Such discussion has enabled us to obtain some clear insights into the problem. Using this knowledge, it is now possible to formulate some tentative postulates that may assist in bringing about a measure of regional closure in the developing countries. It would be too much of a simplification, if not superficiality in thinking, to pretend that we are presenting any invincible solutions to the problem. All that we propose to do here is to point to some departures, (based on our theoretical discussion as well as the accumulated evidence of diagnostic studies and pilot experiences relating to backward area development) that seem to be necessary for influencing the closure of the inter-regional gap existing within the developing countries. With this objective, we may now pull the threads of our analysis and recount what these departures can be.

The experience of the developing countries, which tried to imitate the macro level or top-down planning approaches followed by the developed countries, has brought out the utter futility of such aproaches in tackling the regional and inequality problems met with in their situations. These macro level aproaches, as noted earlier, based themselves on concepts of neo-classical economic analysis and relied heavily on large-scale, vertically organised institutions, either public or private. They also tended to rely excessively on market mechanisms and formal institutional processes which naturally tended to emphasize economic achievements and material aspects of growth, that is, the efficient criteria of investment planning.[3] The development experience has also revealed that these approaches tend to "fly over and by-pass" the poor people or those regions described as disadvantaged or backward. In order that this situation may be corrected, the macro-level approach must be combined with a micro-level approach with its thrust on the local resources and problems of the disadvantaged groups. This implies that planning should come nearer to the grass roots, that is, to planning from below and increasingly area-specific and people-specific.

Strategies for spatial equalization require not merely approaches to planning from below, but also planning at macro-levels. This is because regional economies within any country are not closed economies, and the conditions, efforts and decisions external to the region have also a significant impact on its development. At the same time, our analysis has indicated clearly that in any approach that may be devised for the development of backward regions, the problem of leakages must be effectively countered. The reduction of leakages would mean that development should be *internalized*, that is, any surplus generated should be recycled within the regional boundary. Building those elements and forces that would assist in internalizing development within a backward area then seems to be an essential goal of the development strategy for backward areas. Obviously, the entire development cannot be internalised within an area, as any area is but a sub-system of the national economy and would be characterised by closure for some aspects and would be open in regard to others. Hence what is called for is a strategy of *selective spatial closure*. Such closure has to be both in terms of a 'spatial closure', as well as a certain target group closure", as leakages are possible both from the backward to the developed areas and from the poorer to the richer sections of the community. Therefore, what we would like to emphasize here is that the planning of any backward region within a nation constitutes a typical case of control of an open system. From an operational standpoint, the degree of autonomy left to a backward area such a unit for planning with respect to the setting of ends and the choice of means for planning is bound to be limited, as the national imperatives will already be there and cannot be ignored. In this context, the kind and degree of directional departures that can be exercised in a given situation would call for the exercise of considerable planning skills and talents. The broad parameters have been, by and large, externally defined, some manoeuvrability may still lie in regional or local strategy formulation and in selection of priorities. This limited scope available for manipulation by the planner has to be grasped. The degree of success or failure in backward areas development will therefore depend on this understanding and the extent to which the strategies and priorities are realistically formulated and defined on the one hand on how the internalizing forces of the economy are effectively brought into play, on the other. In this context, "functional programmes"* that will increasingly release the resource potentials as well as the potentials of existing investments will have to be emphsised, so that development projects would confer increasingly enlarged benefits to the backward area. Rendering programmes increasingly functional would imply making them integrated, and mutually supporting; since the backward areas in the developing countries are predominantly rural, it follows logically that Integrated Rural Development should become the cornerstone of strategy for the development of backward areas. Further, as we noted earlier, since spatial closure and target group closure have to be the twin prongs of this strategy, it would follow that the Integrated Rural Development strategy has to be operated on an intensive area basis combined with target group approaches. It was also noted by us during the course of our discussion, that low productivity, poverty and unemployment are the major elements in the backwardness characterising the developing countries. The increase of

* Examples of functional programmes are : Improvement of technology, Programme of feeder roads or credit programmes. Such progrmmes are effective in stimulating other developments, e g. Demand for fertilizers may be stimulated by profitable technologies; Production for market may be prompted by feeder roads; organised credit will stimulate the production programmes in agriculture and industry.

productivity and a major reduction, if not total elimination of poverty and unemployment would therefore become the pivotal thrusts in the Integrated Rural Development strategy. However, it should be emphasised that low productivity, poverty and unemployment are but the symptoms of backwardness. The frontal attack on these problems should, therefore, go still deeper, that is, to the very roots of these problems, which are embedded in the socio-economic milieu, in an intricate, closely interwoven and diversely interlinked pattern of associations and relationships. Therefore, growth strategies, anti-poverty strategies and employment strategies would become sub-strategies in the broader strategy of Integrated Rural Area Development. The precise programmes that will be formulated under these strategies, their appropriate mix, priorities, linkages, complementarities and time-phasing will be further areas of concern in development planning. They cannot be generalized too readily, for they will depend on : (i) the actual syndrome of poverty and inequality characterizing the particular regional situation; (ii) the trade off between growth and capital accumulaticn on the one hand and equalization and redistribution on the other during the long run and short run phases respectively. This analysis takes social accounting into account, particularly the social time preference for investment choices.

In regard to the first consideration, the diagnostic phase of planning is important. It would mean the areal characterization of the syndrome of poverty and inequality, which had been earlier described as identifying the typologies of backwardness. The recognition of this areal differentiation of underdevelopment or the backward area typologies within a country, therefore, constitutes an essential step to aid policy determination for the different backward areas in a country. The non-recognition of this vital spatial dimension of underdevelopment has too often led many a developing country to devise ad hoc and blanket programmes on sectoral lines. The necessity to depart from such an approach is now quite clear. The distinctly geographical or spatial strategies applying to underdevelopment will relate not merely to the reasons for the specific nature of backwardness characterizing an area but also to devising the appropriate solutions for alleviating underdevelopment in that particular setting. They converge on three key geographical concepts namely, (i) areal differentiation; (ii) spatial interaction; and (iii) spatial integration. The contributions of geographers, particularly Philbrick[4], Ullman[5] and Larimore[6] must be mentioned in this context. The ideas thrown up by these authors may be summed up as follows : Man's utilization of the earth leads to an areal differentiation, as the culture advances from the most primitive hunting and gathering stages. Some areas are cultivated, others built up, and some developed into roads. Division of labour and specialization[7] are among the main agents of change. In the course of the development process, the economy moves from largely subsistence to largely exchange. This transformation is accompanied by an increase in the interaction (transport, all sorts of communication, exchange of ideas, and so on), between different areas. This process of differentiation leads to continued expansion and intensification of all connections. In this way, the processes of areal differentiation and spatial interaction ultimately lead to spatial integration. In other words, an isolated area of subsistence activities develops ultimately into a complex, differentiated and highly integrated space economy.

Building on these seminal ideas, this author has further elaborated a paradigm[8] of spatial incidence of economic development which seeks to throw light on the process of development of underdevelopment.

This is indeed a paradigm of shifts envisaging changes in various fronts—economic, spatial structural, population, social, political and administrative spheres. The case studies made in the context of the development of underdevelopment have convinced this author that unless the underdeveloped economy moves along certain envisioned lines, it cannot progress smoothly towards the goals of inter-regional equalization. So a stages approach is necessary to progress towards this goal. But these stages are not inevitable or inexorable. They have to be induced and brought about. To make this happen, development has to be viewed as a controlled experiment, atleast for the identified backward or problem areas and such controlled development has to be backed up with political will as well as continuity and consistency in government's policies. The conceptualization attempted here is based on such premises. If

Paradigm of Shifts

Department aspects	Stage I	Stage II	Stage III
1	2	3	4
Economic	Primary production in settled areas	Intensification and diversification of the economy	Towards capturing regional comparative advantage and interlinkages.
	Opening up of new lands and frontier development	Application of science and technology. Diffusion of innovations. End of 'frontier development'	Interdependent development through interregional specialization.
Spatial	Spatial structure characterised by features of colonialism	Changing spatial structures in varying degrees of "dependency syndrome" (Colonial structures persisting through internal colonialism)	Towards spatial integration.
Population	Redistribution of population-internal migration and adjust-ments	Internal migration of different orders	Population adjusted to 'Territorial Division of Labour'
	Low base of skills	Beginnings of skill development.	Acquisition of advanced skills. Intense mobility of labour

1	2	3	4
Social and Political	Economic feudalism and power structure of the elites.	Shifting power structure in community. Growth of people's institutions trends towards social mobilisation	Democratic Decentralisation
	Excessive class, group or ethnic differentiation.	Class, Group or ethnic confrontations Regional claims and conflicts	Integration of Society
Administrative	Centralized administration Top-down structures	Executive decentralisation i.e. De-concentration	Decentralisation with delegation of powers and authority. Towards integration of development administration with democratic institutions.
Nature and extent of inequalities	C. Sharp, exhibiting features of unbalanced use of natural resources (Eco-destruction)	G. Un-balanced development; the need for eco-development (discordance)	G. Towards maximization or regional potentials (Accordance)
	D. Commencement of inequalities	D. Sharp and fluctuating	D. Towards a pattern of "inequilibrium within tolerance limits"
Nature and extent of inequalities	SE-Sharp	SE-Increasing	SE-Declining; Towards an egalitarian society.

1	2	3	4
Process :	Selective differentiation of problems and determining the Areas for priority intervention based on region's comparative advantage.	Expansion (widening of the bases of interaction) and consolidation. Iterative process of differentiation followed by partial integration	Integration

Note : *G* refers to Geographic inequalities; *D*—refers to Development inequalities; *SE*—refers to Socio-economic inequalities.

such assumptions hold good, the directions of change could be on the above hypothetical lines.

The first and second stages in the paradigm of shifts are derived from the development experiences of the five countries studied here. The third stage is conceptual, a shadowy hypothesis. This represents a fine articulation of the space economy in which the spatial pattern will be characterised by perfect complementaries, linkages and integration of functions.

In this whole process of transformation, a temporal sequence of development is implied. Crucial to this entire process of change is technology. The region absorbs or repels the technological impacts in varying intensities, according to whether these are consistent or inconsistent with its existing form and stage of development. In the temporal sequence of development that we have envisaged, technology comes into play at the very commencement of the transition phase and itself undergoes transformation in qualitative content with the passage of time, as the region will be in a position to absorb progressively higher levels of technology, as development proceeds.

An important question is : how are we to induce this developmental sequence? Here some methods are involved which are indicated against process in the above paradigm. The conceptualization regarding these methods is based on the insights derived from some Indian experiences in backward area development. (There are also some parallels in the Thailand Plan documents). These methods which follow a Point-Line-Network principle, are indicated below against the three phases of transformation envisaged.

Development of Backward Areas—A Conceptualization

PHASE I DIFFERENTIATION PHASE	Limited Intervention (Selectivity)	
	Deep Thrust into Selected Problem Area(s) Cracking Down the Initial Barriers	POINT
PHASE II	Expansion of Coverage	

EXPANSION/ CONSOLIDATION PHASE	Project and Programme Integration Consolidation with Allied Programmes and Activities Iterative process of differentiation followed by limited integration.	LINE
PHASE III INTEGRATION PHASE	Sectoral and Spatial Integration Administrative Coordination Comprehensive Planning	NETWORK

There can be no deterministic planning for the backward areas. Planning interventions can only give some directions of change, both economic and social. These directions have to be absorbed by human beings and the interventions have to be channelized through the various institutional mechanisms available to produce their desired results. Therefore, the cultural and institutional environment in which the strategies and policies are to be implemented must deserve careful attention. In many backward areas, the indigenous and institutional capabilities for absorbing development have to be built up simultaneously, as the programmes are put through. In the developing countries, however, this will only be a gradual process.

Historical experience proves that the development of under-development is neither regular nor cumulative. On the contrary, it is jerky and made up of phases of extremely rapid economic growth followed by sudden blockages and failure to take off. The blocking of transition could be prevented only by strengthening the autocentric and autodynamic forces of development. It is in this context that the self-reliant approach to development involving planning from below, leading to the integral mobilization of local and regional, natural, human, institutional and other resources are stressed. "Capillary mechanisms" are essential to provide the desired orientation to development, for example, bias towards specific target groups.

The bottom-up approach of the kind advocated here, involves multiple objectives such as more production, more employment, better distribution of income and elimination of poverty.

It is not easy to synthesize these objectives. While seeking to operationalize these ideas, apart from the problem of trade off among the different objectives, and the investment allocation problems which they entail, several other obstacles of an administrative and political nature will also have to be taken into account. For example, the success of the programme would depend upon the degree of decentralization of decision-making powers to the various local units; but the central and the State governments may not be willing to effect such complete decentralization. The administrative system may have to be adapted and suitably re-oriented towards the integrated functioning of the rural programmes; where necessary, new institutions will have to be created, but such a reorganization and institutional development may not take place simultaneously. The remodelling of the administrative machinery of an area is always a lengthy process, which needs continual revival to adapt itself to the strategy. Further, since the idea is to encourage self-reliant development by increasing the competitive position of the weaker local producers, considerable support measures as well as policies intended to eliminate penetration by international and extra-regional forces into the regional economy have to be planned as part of the bottom-up development process. Allocation of adequate financial resources for undertaking local development and placing such development on a self-reliant footing, re-distributing assets and income and providing equal access to the publicly created assets to all social classes will be other measures. This kind of socio-structural change visualized here would call for a strong political commitment towards these objectives. All these tasks are, of course, difficult to carry out and make the bottom-up development process an uphill task; but the difficulties should not be a reason for abandoning the effort, as it is now clearly recognized that this is the only approach which is relevant to the developing countries. Bottom-up approaches cannot be practised in isolation. What is desirable is a combination of top-down and bottom-up approaches, mutually supplementing each other and woven into a multi-level frame-work. Undoubtedly, this is a long road for the developing countries discussed here.

While the conceptualization of progression towards the stages of development indicated in the paradigm of shifts outlined above, has followed a certain logic and has been comparatively easy to visualize, the policy-mix desired to reach these stages cannot be visualized in this manner. It will depend on the political and social time preferences on the one hand, and the desired magnitude of gap narrowing envisaged on the other. The policy package desirable for each country will therefore be varying. When it comes to generalization in multi-national comparative situations, we have to lift our sights and ignore the differences or unique characteristics distinguishing specific country situations and recognize only some universalities. If that is so, we may offer only a catalogue of insights relating to specific strategies and the dominant lines of policies. These originate or derive from major problems diagnosed in the particular backward area situation. As mentioned earlier, their mix and their social and political time preferences are matters for the individual countries to decide in the context of their own present and prospective ideological and systemic frameworks. The presentation of this catalogue of insights in the tabulation form below follows the linkage analysis logic.

Major causes of underdevelopment	Strategies called for	Dominant lines of policy
1	2	3
1. A petrified spatial structure historically evolved.	Bottom up strategies involving integral mobilization of local and regional, natural, human, institutional resources of the area; spatial strategies of diffusion/decentralization	Integrated rural development on area basis; multi-level planning, growth centre policies; selective capital intensive investment in projects with strong linkage effects.
2. Leakage of developmental effects	Strategies of selective spatial closure as well as selective target groups closure.	Blocking extra-regional exploitative mechanisms; selective incentive measures applicable only to backward areas and selected backward communities; intensive micro-level planning and development based on local resources.

1	2	3
3. Low productivity	Growth-oriented strategies	Incentives for production based programmes on a selective basis; improvement of technologies; provision of extension services and linkage mechanisms for production; strengthening the delivery systems in the rural areas.
4. Poverty	Anti-proverty strategies	Redistribution of wealth, assets and output as well as allocation of productive resources in favour of the poor, consumption transfers; improved access to land, education, credit facilities and public services for the poor; reorientation of production and demand towards labour intensive, mass consumption goods; target group-oriented programmes; basic needs programme.
5. Unemployment	Employment-oriented strategies	Creation of diversity of employment opportunities through rural development programmes including non-agricultural activities and asset creating public works; improving skills, wage levels job stability and job mobility; building up entrepreneurship.
6. Absence of effective vertical and horizontal linkages	Integrated development strategies; spatial interaction strategies	Improving coordination mechanisms; establishing forward and backward project linkages; improving intra-regional as well as inter-regional accessibility.
7. Elitism/socio-political constraints	Social mobilization strategies	Improving capacities for local and regional self-governing organisations; improving opportunity structures for the disadvantaged groups; conscientisation policies to make the marginalized groups

1	2	3
		aware of their rights and responsibilities.
8. Social/ethnic	Strategy of incrementalism. Integration and assimilation strategies	Point-Line Network policies; protective policies in combination with other productive and target-group oriented policies and programmes.

Note : It should be noted that the various strategies and policy-measures indicated in the above table are nct intended to be use in isolated form but in appropriate combinations to reinforce the total effect.

The problem of inter-regional closure—or narrowing the gap between the advanced regions and the backward regions in a country—is admittedly a hard path. The evolutionary changes in our concepts and thinking about the development process over the years has now enabled us to adjust our sights so that the problems are now seen more clearly and realistically. In keeping with such changing perceptions, it is necessary that our policy needs should also be revised suitably. Many developing countries today are already experimenting with one or more ingredients of this approach. The integrated rural development and block development approach in India, the Kabupaten programme of Indonesia, the Rural Development Programme of Malaysia, the Agro-Ville approach in Pakistan, the Village Production Committee Approach of Sri Lanka, the Village Development Programme and Accelerated Development Programme of Thailand may be mentioned as examples in this direction. The literature emerging from the developing countries[9] reveal clearly that they are all still experiments and do not contain adequate mechanisms for internalising the development process and to bring about inter-regional closure to the required degree of satisfaction. The author is convinced that while seeking a solution to this intractable problem, it is not the rate of growth that has to be accelerated; it is the direction that has to be changed. This process of change has to be vigorously and continuously pursued to its logical end through different stages in a development continuum in a repetitive process alternating between differentiation and integration approaches. This is our main thesis in this work. It is the contention of the author that it would be quite feasible to strike a balance between growth and equity objectives in development and keep spatial inequalities in these countries within tolerance limits. But it would require very substantial modifications in both national and international policies. These are, however, unlikely to take place without a considerable reordering of economic and social priorities in these countries and a restructuring of their administrative apparatus. Above all political will and continuity in policies are necessary.

Undoubtedly, the developing countries have miles and miles to go "before tackling successfully this highly complex, immensely diverse and indeed intractable problem of inequalities; but our contention is that even a single step towards improvement in this direction over the previous position is not a neglible accomplishment. As Galbraith puts it : "the basic comparison in economic affairs is with the position of very near neighbours and the year before".[10]

NOTES AND REFERENCES

1. A.E. Gilbert (Ed), *Development Planning and Spatial Structure*, John Wiley, 1976.
2 *Friedmann* : *Regional Development Policy: A Case Study of Venezuela*, Cambridge, Massachusetts, M. I. T. Press, 1966; W. Alonso, 'Urban and regional imbalances in economic development in *Economic Development and Cultural Change*, Vol, 17, 1969, pp. 1-14.
   (i) L. Rodwin, *Nations and Cities*, Houghton Miffin, New York, 1970.
   (ii) B. J. L. Berry, Relationship between regional economic development and the urban system : The Case of Chile; *Tijdschrift Voor Economische and Social Geografie*, Vol. 60, 1979, pp. 283-307.
3. According to Stohr and Todtling, "The top-down approach has been necessarily stereotyped, not functional in relation to the region's real requirements and consequently socially disruptive" see Stohr, W. and F. Todtling, Spatial equity—Some Antitheses to current regional development doctrine", Paper presented at the 17th European Regional Science Congress, Cracow, 1977.
4. A. Philbrick, Principles of areal functional organisation in regional human geography, *Economic Geography*, Vol. 33, 1957, pp. 299-336.
5. E. L. Ullman, Geographic theory and underdeveloped areas in Ginsburg, N. (Ed.), *Essays on Geography and Economic Development*, Chicago, Department of Geography, Research Paper No. 62, 1960, pp. 26-32.
6. A. Larimore : A measure of economic change: Sequent development of occupance in Busoga District, Uganda—in Ginsburg (ed.) Ibid pp. 111-123.

7. Philbrick terms this process geographic division of labour and Bhuchanan 'specialisation of areas.' The Russian geographers call this 'territorial division of labour.' See A. K. Philbrick, *This Human World*, John Wiley, New York.
8. K. V. Sundaram, Regional planning and tribal development in *Urban and Regional Planning in India*, New Delhi, Vikas, 1977, pp. 281-298.
9. UN ESCAP; The Practice of Local Level Planning—Case Studies in selected rural areas in India, Nepal and Malaysia 1980, Bangkok and Thailand.
10. J. K. Galbraith : *The Underdeveloped Country*, Ibid, p. 45.

INDEX